Application of Liquid Chromatography in Food Analysis

Application of Liquid Chromatography in Food Analysis

Editors
Oscar Núñez
Paolo Lucci

MDPI • Basel • Beijing • Wuhan • Barcelona • Belgrade • Manchester • Tokyo • Cluj • Tianjin

Editors
Oscar Núñez
University of Barcelona (INSA-UB)
Spain

Paolo Lucci
University of Udine — UNIUD
Italy

Editorial Office
MDPI
St. Alban-Anlage 66
4052 Basel, Switzerland

This is a reprint of articles from the Special Issue published online in the open access journal *Foods* (ISSN 2304-8158) (available at: https://www.mdpi.com/journal/foods/special_issues/liquid_chromatography).

For citation purposes, cite each article independently as indicated on the article page online and as indicated below:

LastName, A.A.; LastName, B.B.; LastName, C.C. Article Title. *Journal Name* **Year**, *Article Number*, Page Range.

ISBN 978-3-03943-362-9 (Hbk)
ISBN 978-3-03943-363-6 (PDF)

© 2020 by the authors. Articles in this book are Open Access and distributed under the Creative Commons Attribution (CC BY) license, which allows users to download, copy and build upon published articles, as long as the author and publisher are properly credited, which ensures maximum dissemination and a wider impact of our publications.

The book as a whole is distributed by MDPI under the terms and conditions of the Creative Commons license CC BY-NC-ND.

Contents

About the Editors vii

Oscar Núñez and Paolo Lucci
Application of Liquid Chromatography in Food Analysis
Reprinted from: *Foods* **2020**, *9*, 1277, doi:10.3390/foods9091277 1

Carolina Cortés-Herrera, Graciela Artavia, Astrid Leiva and Fabio Granados-Chinchilla
Liquid Chromatography Analysis of Common Nutritional Components, in Feed and Food
Reprinted from: *Foods* **2019**, *8*, 1, doi:10.3390/foods8010001 5

Monica Rosa Loizzo, Paolo Lucci, Oscar Núñez, Rosa Tundis, Michele Balzano, Natale Giuseppe Frega, Lanfranco Conte, Sabrina Moret, Daria Filatova, Encarnación Moyano and Deborah Pacetti
Native Colombian Fruits and Their by-Products: Phenolic Profile, Antioxidant Activity and Hypoglycaemic Potential
Reprinted from: *Foods* **2019**, *8*, 89, doi:10.3390/foods8030089 67

Md Obyedul Kalam Azad, Won Woo Kim, Cheol Ho Park and Dong Ha Cho
Effect of Artificial LED Light and Far Infrared Irradiation on Phenolic Compound, Isoflavones and Antioxidant Capacity in Soybean (*Glycine max* L.) Sprout
Reprinted from: *Foods* **2018**, *7*, 174, doi:10.3390/foods7100174 77

Anaïs Izquierdo-Llopart and Javier Saurina
Characterization of Sparkling Wines According to Polyphenolic Profiles Obtained by HPLC-UV/Vis and Principal Component Analysis
Reprinted from: *Foods* **2019**, *8*, 22, doi:10.3390/foods8010022 87

Valentina Santoro, Federica Dal Bello, Riccardo Aigotti, Daniela Gastaldi, Francesco Romaniello, Emanuele Forte, Martina Magni, Claudio Baiocchi and Claudio Medana
Characterization and Determination of Interesterification Markers (Triacylglycerol Regioisomers) in Confectionery Oils by Liquid Chromatography-Mass Spectrometry
Reprinted from: *Foods* **2018**, *7*, 23, doi:10.3390/foods7020023 97

Núria Carranco, Mireia Farrés-Cebrián, Javier Saurina and Oscar Núñez
Authentication and Quantitation of Fraud in Extra Virgin Olive Oils Based on HPLC-UV Fingerprinting and Multivariate Calibration
Reprinted from: *Foods* **2018**, *7*, 44, doi:10.3390/foods7040044 107

About the Editors

Oscar Núñez studied chemistry at the University of Barcelona where he also received his Ph.D. in 2004. He worked as visiting researcher for half a year in the development of on-line pre-concentration methods using micellar electrokinetic chromatography (MEKC) at the University of Hyogo (Japan) in collaboration with Professor Terabe (father of MEKC technique). Starting in October 2005, he joined the Kyoto Institute of Technology (Japan) as a two-year post-doc researcher working with Professor Tanaka developing monolithic silica capillary columns under a fellowship from the Japan Society for the Promotion of Science. In November 2007, he joined the University of Barcelona again, working in the Chromatography, Capillary Electrophoresis and Mass Spectrometry research group. Since December 2014, he has been a Serra Húnter Professor at the Section of Analytical Chemistry (Department of Chemical Engineering and Analytical Chemistry, University of Barcelona). He has more than 130 scientific papers and book chapters to his name, and he is editor of three books on LC-MS/MS, sample preparation techniques in food analysis, and capillary electrophoresis. He has extensive experience in the development of liquid chromatography methods coupled to low- and high-resolution mass spectrometry, as well as sample treatment procedures, for environmental and food analysis. His major research areas nowadays involve the characterization, classification and authentication of food and natural products, as well as the prevention of food frauds.

Paolo Lucci attained his Ph.D. in 2008 at the SAIFET department of the Polytechnic University of Marche (Italy). In February 2009, he joined the NASCENT European project as an experienced researcher at POLYIntell SAS (France) and then in 2010 he spent one year as an experienced researcher at the Department of Analytical Chemistry of the University of Barcelona within the Carbosorb European Project. In April 2011, he joined the School of Sciences of the Pontificia Universidad Javeriana (Colombia) where he was named head of the research group of "Foods, Nutrition and Helath" in 2012, and then head of the Department of Nutrition and Biochemistry in 2014. Currently, he is a Senior Researcher at the Department of Agri-Food, Environmental and Animal Sciences at the University of Udine (Italy). He has published more than 70 scientific papers and book chapters, and he is editor of two books focused on LC-MS and sample preparation techniques in food analysis. He has extensive experience in liquid chromatography coupled with tandem mass spectrometry (LC-MS/MS), molecularly imprinted polymers (MIPs), as well as alternative treatment procedures for environmental and food sample analysis. Other research areas include the chemical characterization of natural extracts and the evaluation of their health-promoting effect through in vitro and in vivo studies.

Editorial

Application of Liquid Chromatography in Food Analysis

Oscar Núñez [1,2,3,*] and Paolo Lucci [4,*]

1. Department of Chemical Engineering and Analytical Chemistry, University of Barcelona, Martí i Franquès 1-11, E08028 Barcelona, Spain
2. Research Institute in Food Nutrition and Food Safety, University of Barcelona, Recinte Torribera, Av. Prat de la Riba 171, Edifici de Recerca (Gaudí), Santa Coloma de Gramenet, E08921 Barcelona, Spain
3. Serra Húnter Fellow, Generalitat de Catalunya, Rambla de Catalunya 19-21, E08007 Barcelona, Spain
4. Department of Agri-Food, Environmental and Animal Sciences, University of Udine, 33100 Udine, Italy
* Correspondence: oscar.nunez@ub.edu (O.N.); paolo.lucci@uniud.it (P.L.)

Received: 9 September 2020; Accepted: 10 September 2020; Published: 11 September 2020

Food products are very complex mixtures consisting of naturally-occurring compounds and other substances, generally originating from technological processes, agrochemical treatments, or packaging materials. Several of these compounds (e.g., veterinary drugs, pesticides, mycotoxins, etc.) are of particular concern because, although they are generally present in very small amounts, they are nonetheless often dangerous to human health. On the other hand, food is no longer just a biological necessity for survival. Society, in general, demands healthy and safe food, but it is also increasingly interested in other quality attributes more related to the origin of the food, the agricultural production processes used, the presence or not of functional compounds, etc. In an increasingly populated world and with an increasingly demanding society regarding food quality, food production has become a completely global aspect on a global level. In addition, in this field, where the number of people involved in the food production process, from its origin to its consumption, is enormous, it is increasingly difficult to guarantee the integrity and, above all, the authenticity of foodstuffs. In consequence, improved methods for the determination of authenticity, standardization, and efficacy of nutritional properties in natural food products are required to guarantee their quality and for the growth and regulation of the market. Thus, food safety and food authentication are hot topics for both society and the food industry.

Nowadays, liquid chromatography with ultraviolet (LC-UV) detection, or coupled to mass spectrometry (LC-MS) and high-resolution mass spectrometry (LC-HRMS), are among the most powerful techniques to address food safety issues and to guarantee food authenticity in order to prevent fraud [1–8]. The aim of this Special Issue "Application of Liquid Chromatography in Food Analysis" was to gather review articles and original research papers focused on the development of analytical techniques based on liquid chromatography for the analysis of food. This Special Issue is comprised of six valuable scientific contributions, including five original research manuscripts and one review article, dealing with the employment of liquid chromatography techniques for the characterization and analysis of feed and food, including fruits, extra virgin olive oils, confectionery oils, sparkling wines and soybeans.

Cortés-Herrera et al. reviewed the potential of liquid chromatography for the analysis of common nutritional components in feed and food [9]. Food and feed share several similarities when facing the implementation of liquid-chromatographic analysis. Using the experience acquired over the years through the application chemistry in food and feed research, the authors selected and discussed analytes of relevance for both areas. This interesting review addresses the common obstacles and peculiarities that each analyte offers for the implementation of LC methods throughout the different steps of the method development (sample preparation, chromatographic separation and detection). The manuscript consists mainly of three sections: feed analysis (at the beginning of the food chain); food destined for human consumption determinations (the end of the food chain); and finally, assays shared by either matrices or laboratories. Polyphenols, capsaicinoids, theobromine and caffeine, cholesterol, mycotoxins, antibiotics, amino acids, triphenylmethane dyes, nitrates/nitrites, ethanol soluble carbohydrates/sugars, organic acids, carotenoids, and hydro and liposoluble vitamins are examined.

Several original research works reported the application of liquid chromatography-based analytical methodologies for the characterization of food products. Loizzo et al. characterized native Colombian fruits and their by-products by determining their phenolic profile, antioxidant activity and hypoglycaemic potential [10]. The use of ultra-high performance liquid chromatography-high resolution mass spectrometry (UHPLC-HRMS) with an Orbitrap mass analyzer revealed the presence of chlorogenic acid as dominant compound in Solanaceae samples. In addition, and based on the Relative Antioxidant Score (RACI) and Global Antioxidant Score (GAS) values, *Solanum quitoense* peel showed the highest antioxidant potential among Solanaceae samples, while *Passiflora tripartita* fruits exhibited the highest antioxidant effects among Passifloraceae samples. Considering that some of the most promising results were obtained by the processing waste portion, the authors highlighted that its use as functional ingredients should be considered for the development of nutraceutical products intended for patients with disturbance of glucose metabolism. Obyedul Kalam Azad et al. evaluated the effect of artificial LED light and far infrared (FIR) irradiation on phenolic compounds, isoflavones and the antioxidant capacity of soybean (*Glycine max* L.) sprouts [11]. Six isoflavones (daidzin, glycitin, genistin, daidzein, glycitein and genistein) were determined by LC. The authors applied artificial blue (470 nm) and green (530 nm) LED and florescent light (control) on soybean sprouts, from three to seven days after sowing in the growth chamber. Total phenolic content, antioxidant capacity and total isoflavones content were higher under blue LED compared to control. Thus, results suggested that blue LED was the most suitable light to steady accumulation of secondary metabolites in growing soybean sprouts. In another interesting research work, polyphenolic profiles obtained by high-performance liquid chromatography with ultraviolet/visible detection (HPLC-UV/Vis) and principal component analysis (PCA) were employed by Izquierdo-Llopart and Saurina for the characterization of sparkling wines [12]. Chromatographic profiles were recorded at 280, 310 and 370 nm to gain information on the composition of benzoic acids, hydroxycinnamic acids and flavonoids, respectively. The authors employed the obtained HPLC-UV/vis data, consisting of composition profiles of relevant analytes, to characterize cava wines produced from different base wine blends by using chemometrics. Other oenological variables, such as vintage, aging or malolactic fermentation, were fixed over all the samples to avoid their influence on the description. PCA and other statistic methods were able to extract the underlying information and provided an excellent discrimination of the analyzed samples according to their grape varieties and coupages. Finally, Santoro et al. performed the characterization and determination of interesterification markers (triacylglycerol regioisomers) in confectionery oils by liquid chromatography-high resolution mass spectrometry (LC-HRMS) [13]. In the confectionery industry, controlling the formation degree of positional isomers is important in order to obtain fats with the desired properties. The separation of triacylglycerol regioisomers is a challenge when the number of double bonds is the same and the only difference is in their position within the triglyceride molecule. The authors aimed to obtain a chromatographic resolution that might allow reliable qualitative and quantitative evaluation of triacylglycerol positional isomers within rapid retention times, and robustness in respect of repeatability and reproducibility by means of LC-HRMS using an LTQ-Orbitrap analyzer with atmospheric pressure chemical ionizacion (APCI). The time required for the global analysis was relatively short, the chromatographic resolution and efficiency were satisfactory and the mass detection allowed for identifying the isobaric components of each position isomer couple. In conclusion, the described method may well be considered a good diagnostic tool of interesterification consequences that are strictly connected to confectionery product quality.

HPLC-UV was also proposed by Carranco et al. for the authentication and quantitation of frauds in extra virgin olive oils (EVOOs) [14]. For that purpose, HPLC-UV chromatographic fingerprints recorded at 257, 280 and 316 nm were employed as sample chemical descriptors for the characterization and authentication of monovarietal EVOOs and other vegetable oils by chemometrics. PCA results showed a noticeable discrimination between olive oils and other vegetable oils using raw HPLC-UV fingerprints as data descriptors. However, the authors observed that selected HPLC-UV chromatographic time-window segments were able to improve the discrimination among the monovarietal EVOOs analyzed. In addition, partial least squares (PLS) regression was employed to tackle olive oil authentication of Arbequina EVOO adulterated with Picual EVOO, refined olive oil, and sunflower oil, achieving highly satisfactory results with overall errors in the quantitation of adulteration in the Arbequina EVOO (minimum 2.5% of adulterant) below 2.9%.

In summary, the Special Issue "Application of Liquid Chromatography in Food Analysis" demonstrated the great importance of liquid chromatography analytical methodologies to address the characterization and determination of targeted compounds in food, as well as to guarantee food integrity and authenticity. In addition, several authors have also demonstrated the requirement of advanced chemometric approaches as essential tools in combination with LC to achieve robust results, not only with the objective of characterizing food composition, but also to obtain satisfactory classification methods, and to prevent food fraud.

Author Contributions: Writing—original draft preparation, O.N.; writing—review and editing, O.N. and P.L.; All authors have read and agreed to the published version of the manuscript.

Funding: This research received no external funding.

Conflicts of Interest: The authors declare no conflict of interest.

References

1. Malik, A.K.; Blasco, C.; Picó, Y. Liquid chromatography-mass spectrometry in food safety. *J. Chromatogr. A* **2010**, *1217*, 4018–4040. [CrossRef] [PubMed]
2. Di Stefano, V.; Avellone, G.; Bongiorno, D.; Cunsolo, V.; Muccilli, V.; Sforza, S.; Dossena, A.; Drahos, L.; Vékey, K. Applications of liquid chromatography-mass spectrometry for food analysis. *J. Chromatogr. A* **2012**, *1259*, 74–85. [CrossRef] [PubMed]
3. Núñez, O.; Gallart-Ayala, H.; Martins, C.P.B.; Lucci, P. New trends in fast liquid chromatography for food and environmental analysis. *J. Chromatogr. A* **2012**, *1228*, 298–323. [CrossRef] [PubMed]
4. Núñez, O.; Gallart-Ayala, H.; Martins, C.P.B.; Lucci, P. *Fast Liquid Chromatography-Mass Spectrometry Methods in Food and Environmental Analysis*; Núñez, O., Gallart-Ayala, H., Martins, C.P.B., Lucci, P., Eds.; Imperial College Press: London, UK, 2015; ISBN 9-781783-264-933.
5. Preti, R. Core-shell columns in high-performance liquid chromatography: Food analysis applications. *Int. J. Anal. Chem.* **2016**, *2016*, 3189724. [CrossRef] [PubMed]
6. Sentellas, S.; Núñez, O.; Saurina, J. Recent Advances in the Determination of Biogenic Amines in Food Samples by (U)HPLC. *J. Agric. Food Chem.* **2016**, *64*, 7667–7678. [CrossRef] [PubMed]
7. Lucci, P.; Saurina, J.; Núñez, O. Trends in LC-MS and LC-HRMS analysis and characterization of polyphenols in food. *TrAC—Trends Anal. Chem.* **2017**, *88*, 1–24. [CrossRef]
8. Esteki, M.; Shahsavari, Z.; Simal-Gandara, J. Food identification by high performance liquid chromatography fingerprinting and mathematical processing. *Food Res. Int.* **2019**, *122*, 303–317. [CrossRef] [PubMed]
9. Cortés-Herrera, C.; Artavia, G.; Leiva, A.; Granados-Chinchilla, F. Liquid chromatography analysis of common nutritional components, in feed and food. *Foods* **2019**, *8*, 1. [CrossRef] [PubMed]
10. Loizzo, M.R.; Lucci, P.; Núñez, O.; Tundis, R.; Balzano, M.; Frega, N.G.; Conte, L.; Moret, S.; Filatova, D.; Moyano, E.; et al. Native Colombian fruits and their by-products: Phenolic profile, antioxidant activity and hypoglycaemic potential. *Foods* **2019**, *8*, 89. [CrossRef] [PubMed]
11. Obyedul Kalam Azad, M.; Kim, W.W.; Park, C.H.; Cho, D.H. Effect of artificial LED light and far infrared irradiation on phenolic compound, isoflavones and antioxidant capacity in soybean (*Glycine max* L.) sprout. *Foods* **2018**, *7*, 174. [CrossRef] [PubMed]
12. Izquierdo-Llopart, A.; Saurina, J. Characterization of sparkling wines according to polyphenolic profiles obtained by HPLC-UV/Vis and principal component analysis. *Foods* **2019**, *8*, 22. [CrossRef] [PubMed]

13. Santoro, V.; Dal Bello, F.; Aigotti, R.; Gastaldi, D.; Romaniello, F.; Forte, E.; Magni, M.; Baiocchi, C.; Medana, C. Characterization and determination of interesterification markers (triacylglycerol regioisomers) in confectionery oils by liquid chromatography-mass spectrometry. *Foods* **2018**, *7*, 23. [CrossRef] [PubMed]
14. Carranco, N.; Farrés-Cebrián, M.; Saurina, J.; Núñez, O. Authentication and quantitation of fraud in extra virgin olive oils based on HPLC-UV fingerprinting and multivariate calibration. *Foods* **2018**, *7*, 44. [CrossRef] [PubMed]

© 2020 by the authors. Licensee MDPI, Basel, Switzerland. This article is an open access article distributed under the terms and conditions of the Creative Commons Attribution (CC BY) license (http://creativecommons.org/licenses/by/4.0/).

Review

Liquid Chromatography Analysis of Common Nutritional Components, in Feed and Food

Carolina Cortés-Herrera [1], Graciela Artavia [1], Astrid Leiva [2] and Fabio Granados-Chinchilla [2,*]

1. Centro Nacional de Ciencia y Tecnología de Alimentos (CITA), Universidad de Costa Rica, Ciudad Universitaria Rodrigo Facio 11501-2060, Costa Rica; carolina.cortesherrera@ucr.ac.cr (C.C.-H.); graciela.artavia@ucr.ac.cr (G.A.)
2. Centro de Investigación en Nutrición Animal, Universidad de Costa Rica, Ciudad Universitaria Rodrigo 11501-2060, Costa Rica; astrid.leiva@ucr.ac.cr
* Correspondence: fabio.granados@ucr.ac.cr; Tel.: +506-2511-2028; Fax: +506-2234-2415

Received: 14 September 2018; Accepted: 5 November 2018; Published: 20 December 2018

Abstract: Food and feed laboratories share several similarities when facing the implementation of liquid-chromatographic analysis. Using the experience acquired over the years, through application chemistry in food and feed research, selected analytes of relevance for both areas were discussed. This review focused on the common obstacles and peculiarities that each analyte offers (during the sample treatment or the chromatographic separation) throughout the implementation of said methods. A brief description of the techniques which we considered to be more pertinent, commonly used to assay such analytes is provided, including approaches using commonly available detectors (especially in starter labs) as well as mass detection. This manuscript consists of three sections: feed analysis (as the start of the food chain); food destined for human consumption determinations (the end of the food chain); and finally, assays shared by either matrices or laboratories. Analytes discussed consist of both those considered undesirable substances, contaminants, additives, and those related to nutritional quality. Our review is comprised of the examination of polyphenols, capsaicinoids, theobromine and caffeine, cholesterol, mycotoxins, antibiotics, amino acids, triphenylmethane dyes, nitrates/nitrites, ethanol soluble carbohydrates/sugars, organic acids, carotenoids, hydro and liposoluble vitamins. All analytes are currently assayed in our laboratories.

Keywords: food and feed analysis; liquid chromatography; challenges; nutritional analysis; additives; contaminants

1. Introduction

Food and feed analysis are paramount to assess both nutritional quality and safety of commodities. Interconnectivity of food sources [1,2] and new processing techniques [3] make for a more diverse and complex food supply. Legal thresholds have been stipulated that establish acceptable levels for individual chemical additives, residues, and contaminants in products [4,5]. Feed is a paramount target for analysis since it situates at the start of the food chain and poor feed quality can affect the yield on food-producing animals [6]. Understanding the complexities of food safety is the goal of approaches such as One Health [7], Farm-to-Fork [6], or MyToolbox [8]. Furthermore, feed contaminants carryover downstream can reach products such as meat, eggs, and milk (see for example the transference of aflatoxin M_1 from aflatoxin B_1-contaminated feed). Ingredients either destined for food or feed production (e.g., cereals) are among the fundamental constituents for several staple commodities. Other regulations require food and feed labeling to list ingredients relating to the nutritional content [9,10]. All stakeholders involved in the food and feed chain must be able to assess product quality and safety. Hence, it is imperative to rely on techniques that meet several analytical performance parameters. More and more, food and feed analysis methods are based on LC (liquid-chromatography) [11,12], which has proven to be an optimal technology for screening, detection, and quantification of a vast variety of analytes (see Table 1). The reason behind this is related to the molecular affinity between the analyte and also: i. the mobile phase (which is usually a mixture of solvents) ii. stationary phase (modified silica and polymer scaffolds). Within the LC approach itself, several alternatives are available for a researcher to resolve a specific task at hand. Each analyte presents its own unique trials.

Table 1. Typical food and feed analytes assayed using HPLC (High-Performance Liquid-Chromatography).

Analyte Category Examples	Relevance in Feed and Food Quality
Additives	
Acidulants Acetic acid, lactic acid, and citric acid [13].	Used in beverage, food, and feed production, are part of the primary metabolism, are often produced by fermentation. Acidic additives serve as buffers to regulate acidity, antioxidants, preservatives, flavor enhancers, and sequestrants. Related to beneficial effects on animal health and growth performance as feed additives.
Antioxidants Gallic, rosemarinic, canosic, and caffeic acids, glabrene, procyanidins, quercetin, catechin α-, β-, γ-, and δ-tocopherols, Eugenol, Carnosine, Tyr-Phe-Glu, and Tyr-Ser-Thr-Ala.	Lipid and protein oxidation can impact meat quality, nutrition, safety, and organoleptic properties. Antioxidants are added during animal production and meat processing to enhance the nutritional and health benefits of meat and minimize the formation of carcinogens for the chemical safety of cooked and processed meats [14,15]. They can also be used to extend food [16] and feed [17] shelf life.
Preservatives Acetates, bacteriocins, benzoates (p-hydroxybenzoic acid), borates, carbonates, lactates, nitrates/nitrites, parabens, propionates, sorbates, and sulfites.	Usually, act as bacteriostatic and bactericidal agents to prevent microbial spoilage, antimicrobials not only extend shelf life, but they also enhance the product's safety [18].
Flavors and fragrances Alcohols, methyl ketones, 2,3-butanedione, lactone, butanoic acid, esters, isovaleric acid, pyrazines, geosmin, vanillin, benzaldehyde, terpenes.	Widely used in food, beverage, feed, cosmetic, detergent, chemical and pharmaceutical formulations [19].
Sweeteners Approved as food additives: saccharin, aspartame, acesulfame potassium, sucralose, neotame, advantame. Generally regarded as safe (GRAS): Steviosides [28–31].	Non-nutritive sweeteners have become an essential part of daily life and are in increasing demand as it is used in a wide variety of dietary and medicinal products [20]. They play a role in the reduction of table sugar [21]. In the case of artificial sweeteners, their use is controversial as they have associated with health risks [20,22] and water pollution [23]; currently, the use of natural sweeteners is supported as an alternative [24]. Sweetened products must be subject to verification to ensure the presence of the sweetener. Furthermore, sweeteners are regulated food additives [25] unless recognized as safe [26,27].
Natural Components	
Analyte Category Examples	Relevance in Feed and Food Quality
Inorganic ions Sulfites, sulfates, phosphate, polyphosphate, nitrate and nitrite, N-nitroso compounds, cyanide, bromide, bromate, chloride, chlorite, fluoride, iodide.	Essential in both raw and processed products, related to food nutritional quality, preservation, technological processing, and safety [32].
Lipids and fatty acids $C_{1:0}$ (formic/methanoic), $C_{2:0}$ (acetic/ethanoic), $C_{3:0}$ (propionic/propanoic), $C_{4:0}$ (butyric/butanoic), $C_{6:0}$ (caproic/hexanoic), $C_{8:0}$ (caprylic/octanoic), $C_{10:0}$ (capric/decanoic), $C_{12:0}$ (lauric/dodecanoic), $C_{12:0}$ (myristic/tetradecanoic), $C_{16:0}$ (palmitic/hexadecanoic), 9c-$C_{16:1}$ (palmitoleic/(9Z)-hexa-dec-9-enoic), stearic, oleic, ricinoleic, vaccenic, linoleic, α-linoleic, γ-linoleic, arachidic, eicosapentaenoic, behenic, erucic, docosahexaenoic, lignoceric, cholesterol [33–37].	Major constituents of foods and feeds, of dietary importance as a significant source of energy. Provide essential fat-soluble nutrients. Are prone to peroxidation. Part of biological membranes.
Biogenic amines Putrescine, histamine, cadaverine.	Nitrogen-based toxic compounds, mainly formed through decarboxylation of amino acids. Relevant for quality and safety of diverse foods such as dairy products [38], fermented goods [39] including wines [40], fishery commodities [41].
Amino acids The main fermentative amino acids for animal nutrition are L-lysine, L-threonine, and L-tryptophan. DL-Methionine.	Part of a protein-containing diet, and as supplemented individual products. Amino acids are used in medical (parenteral) nutrition and dietary supplements [42].
Carbohydrates Glucose is the primary energy source for fetal growth, and lactose is crucial for the development of human and animal infants alike.	The most abundant feed energy in diets for some species of animals [43,44].
Vitamins Fat-soluble: retinol (vitamin A) and retinyl acetate, and palmitate), tocopherols (α- (vitamin E), β-, γ-, and δ- and their acetates), ergocalciferol (vitamin D_2), cholecalciferol (vitamin D_3), phylloquinone (vitamin K_1), menaquinone (vitamin K_2), 7-dehydrocholesterol, β-carotene. Hydrosoluble or B complex vitamins: thiamine (B_1), Riboflavin (B_2), flavin mononucleotide or riboflavin-5'-phosphate, niacin/nicotinamide riboside/niacinamide (B_3), pantothenic acid (B_5), pyridoxine/pyridoxamine/pyridoxal (B_6), biotin (B_7), folates (B_9), cobalamines (B_{12}) [45].	Complex unrelated compounds present in minute amounts in natural foodstuffs. Essential to normal metabolism; their deficiency causes disease.

Table 1. Cont.

Additives	
Analyte Category Examples	Relevance in Feed and Food Quality
Alkaloids Octopamine, synephrine, tyramine, N-methyl-tyramine, hordenine in bitter orange products [47], morphine, codeine, thebaine, papaverine, and noscapine in poppy straw [48], caffeine and trigonelline in coffee [49], indole and oxindole alkaloids in *Unaria* sp. [50], theobromine and caffeine in tea [51] and coffee [52], Harman alkaloids (harmane and harmine) in passion fruit [53], ergot alkaloids in animal feed (ergometrine, ergotamine, ergocornine, ergocryptine, ergocristine [54], piperine [55].	Alkaloids are natural compounds with a characteristic cyclic structure and a nitrogen atom [46]. Alkaloid-containing plants are an essential part of the regular diet, present as natural constituents of several food products [46]. The most common use for alkaloid-containing plants is as stimulants increased concentrations of these compounds can be attained within the food chain as a result of food processing, as food contaminants or as food flavorings [46].

Residues and Contaminants	
Analyte Category Examples	Relevance in Feed and Food Quality
Chemotherapeutics and antiparasitic drugs Tetracyclines [56].	Antibiotics are extensively utilized in productive animals with therapeutic, prophylactic, metaphylactic, growth promoting, and food effectiveness enhancing ends. These practices that have been reflected in veterinary residues in products for human consumption (meat, eggs, and milk) and is also related to directly with allergies and antimicrobial resistance.
Mycotoxins Aflatoxins [58].	Mycotoxins are practically ubiquitous contaminants, classified as teratogenic, carcinogenic and immunosuppresive, and that affects a great variety of grains, fruits and seeds, as well as eggs, dairy products, compounds feeds, and other feed ingredients [57].
Pesticides Atrazine, glyphosate, aminomethylphosphonic acid, phenoxy herbicides.	Used for crop protection and to treat infestations in livestock. Their poor use results in contamination of the environment and the food itself, impacting human health. Residues usually found in vegetables, fruits, honey, fish, eggs, milk, and meat, serving as potential sources of contamination to consumers [59–61].

To successfully analyze or isolate a compound, a researcher is faced with several questions: What is the problem to solve, the objective or purpose for the analysis? Is the required data qualitative or quantitative? Are there two or multiple compounds to be separated? What are the physicochemical characteristics of the target(s)? What matrix was the analyte recovered from and which interferences are expected? What is the amount of analyte expected to be recovered? What equipment is accessible in the laboratory?

Considering the above, a suitable column (Table 2) and detection system must be selected (Table 3). Sample preparation can aid to solve some of these issues, especially those regarding interferences and sensitivity but cannot solve issues with poor detector choice. For example, if sensitivity is a problem using the selected detection system on hand and no other system is available, the initial sample mass can be increased, or a concentration step (evaporation or solid phase extraction (SPE)) can be performed. Additionally, the sample injection volume can be expanded to improve sensitivity.

Table 2. General conditions required for each mode of chromatography.

Type of Liquid Chromatography	Micro	Semi-Micro	Conventional	Semi-Preparative	Preparative	Process
Column internal diameter, mm	$0.3 < x \leq 1.0$	$1.0 < x \leq 3.0$	$4.0 < x \leq 8.0$	$8.0 < x \leq 20.0$	$20.0 < x \leq 50.0$	$x > 50.0$
Eluent flow rate, mL min^{-1}	$0.001 < x \leq 0.1$	$0.1 < x \leq 0.4$	$0.4 < x \leq 2.0$	$2.0 < x \leq 10.0$	$10.0 < x \leq 150.0$	$x > 150.0$

Additionally, automation is relevant for conserving resources and reducing turnover times. An analyst can program an autosampler to increasingly adjust the volume of a standard with a fixed concentration. For example, to construct a calibration curve between 1000 and 62.5 µg L^{-1}, one could use a 1000 µg L^{-1} standard and instruct the sampler to take from the same vial 20 µL, 10 µL, 5 µL, 2.5 µL, 1.25 µL, consecutively. The sampler will construct a calibration curve without analyst intervention and this automation will reduce errors. Autosamplers are designed to inject small volumes without significant loss, with good precision, and adequate reproducibility. They can also inject variable amounts, dilute the sample prior to injection and perform precolumn derivatization [91]. If a sample is outside of calibration standard of higher concentration, an analyst can inject a different volume to ensure it will fit among the calibration curve range. However, injection volume has an impact on peak shape. The method must be validated to show this is a valid approach. (See for example, [92]). Reference for one example of the versatility of an LC system and capabilities for its automation. In this review, we intend to give the reader a thorough background on the common analyses performed, for quality assurance and safety, in food and feed laboratories. We will include the most recent and relevant experience gathered for each test while pointing out the difficulties that each essay presents and the common ground shared by both types of laboratories.

Table 3. Characteristics of the most common detectors used in liquid chromatography.

Detector Type Range of Applications, Attributes, and Minimal Detectable Quantity Limitations	Applications in Feed and Food Quality
Non-Destructive Detectors	
Photodiode-Array (PDA)/Variable wavelength (VW)/UV-vis Selective; universal at low wavelengths, 3D spectra comparison can determine peak purity, can detect nanograms. Chromophore must be preseent, solvents transparent to the wavelength used must be provided.	Sulfonated azo dyes in beverages, hard candy and fish roe samples [62], purity of caffeine reference material [63], sulfamethazine and trimethoprim in liquid feed premixes [64], nitrofurans animal feed [65].
Fluorescence (FL) Very selective and specific; monitors two wavelengths simultaneously, 3D fluorescent spectra, fluorescent fingerprinting/fluorescence pattern analysis [66] gradients do not affect baseline significantly, can detect low picograms. Fluorophore must be present, derivative formation and quenching are often needed.	Sulfonamides [67] and fluoroquinolones [68] in animal feed, aflatoxins in agricultural food crops [69] and milk [70].
Electrochemical (EC) Very selective; oxidation or reduction must be possible, can detect from femtograms to nanograms. Conductive mobile phase, susceptible to background noise and electrode degradation.	Macrolide antibiotics in animal feeds [71], vitamin C in oranges and apples [72].
Refractive index (RI) Universal (All compounds affect refractive properties) and versatile; solvent compatible, relatively simple, can detect micrograms. Gradient incompatible, high S/N ratio when the pump is mixing two or more solvents, susceptible to temperature and flow variation.	Inulin in chicory roots [73], total carbohydrates in wine and wine-like beverages [74].
Conductivity Selective; an ionic form of compound necessary can detect low pictograms. Suppression of mobile phase background conductivity, special equipment required.	Choline, and trimethylamine in feed additives [75], L-carnitine, choline, and metal ions in infant formula [76].
Destructive Detectors	
Mass spectrometry Selective and specific; based on a specified mass/charge ratio, ion fractionation, can detect low nanograms. Expensive, expert users are needed for equipment and data manipulation.	Analysis of acrylamide in food [77], tiamulin, trimethoprim, tylosin, sulfadiazine, and sulfamethazine residues in medicated feed [78], multiclass antibiotics in eggs [79], zearalenone and deoxynivalenol metabolites in milk [80], cattle feed analysis of Aspergillus clavatus mycotoxins [81], choline chloride in feed and feed premixes [82].
Radioactivity Selective; Distribution and mass balance wide response range can detect pictograms. Large-volume flow cells can produce peak broadening and decreased the resolution.	Drug metabolite identification [83].
Evaporative light scattering (ELS) Universal; Nonvolatile analyte nebulization, can detect in the range of nanograms. Volatile buffers required, poor reproducibility and limited dynamic range.	N-acetylglucosamine and N-acetylgalactosamine in dairy foods [84], sucralose and related compounds [85], spectinomycin and associated substances [86].
Corona-Charged aerosol [87] Universal; can detect non-ultraviolet and weakly ultraviolet active compounds [88], ionized particles measured by an electrometer, can detect low nanograms. Volatile buffers required.	Erythritol, xylitol, sorbitol, mannitol, maltitol, fructose, glucose, sucrose, and maltose in food products [89], fumonisins in maize [90].

2. Measurements of Commonly Consumed Food Commodities

2.1. Polyphenols

Polyphenols usually refers to several chemical compounds including flavanols (e.g., catechins and tannins from tea), flavanones (i.e., hesperidin from citrus fruits), flavonols (e.g., quercetin from tea, apples and onions), "chlorogenic acids" (including hydroxycinnamic acids caffeic, ferulic, *p*-coumarinic acids usually extracted from coffee), anthocyanins (which are partly responsible for imparting color to plant structures), and stilbenes (e.g., from berries, grape skins and peanuts) (Figure 1) [93].

These compounds, secondary metabolites from plants [94], have, among other functions, a protective capability within the vegetable tissue, structure, and support [94], and, even, pollinator attraction [95]. For example, chlorogenic acid (i.e., the esterification product between caffeic and quinic acid) is an intermediate in lignin biosynthesis [96]. Data suggests that long-term consumption of such compounds can have beneficial effects [94] as it can improve an organism's antioxidant capacity [93] which in turn relates, for example, to cognitive improvement [97] and reduction in adipogenesis and oxidative stress [98]. Fruits, especially berries, are [97–101] rich in these bioactive compounds, both extractable [102,103] and non-extractable [104]. From the technological standpoint, polyphenol safeguard is paramount to achieve functional foods [105] with added value (e.g., beverages) and a bioactive capacity of compounds as close as those from the raw material. Several operation units have been applied to fruits to assess polyphenol retention after processing including nanofiltration [101], high hydrostatic pressure [106], and drying [107,108]. Method-wise, the solvent has a profound effect on the number and type of polyphenols extracted. Polyphenol analysis must first identify the type of matrix to be analyzed, the chemical nature of the polyphenols of interest, and different solvents and solvent systems should be examined. The most appropriate solvent for the case in hand (i.e., maximizing compound diversity and yield) should be the one selected [109]. For example, Flores and coworkers resuspended the methanolic extract in hexane, chloroform, ethyl acetate, and n-butanol and reanalyzed each fraction. Ethyl acetate fraction exhibited the best results [110]. Finally, though polyphenols are usually related to health applications [111,112], antinutritional effects should be considered [109]. Some examples of polyphenol analysis are included in Table 4.

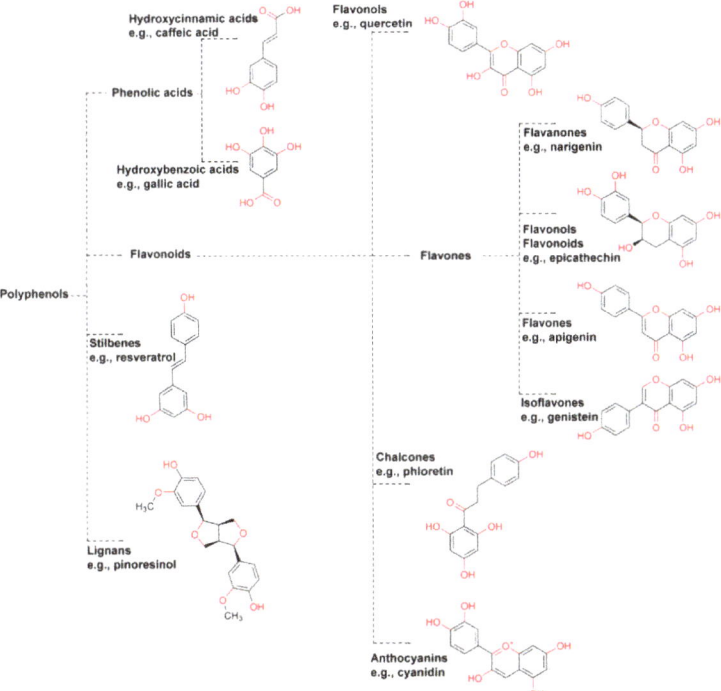

Figure 1. Polyphenols structure and classification [97]. Highly functionalized structures account for the molecules radical scavenging, metal ion chelating, and enzyme inhibition. Hydrogen bonding can stabilize phenoxyl radicals.

Table 4. Polyphenol analysis in different matrices, based on liquid chromatography, and varied approaches to determine them.

Matrix	Analytes Identified	Extraction Method	Measurement Method, Chromatographic Column	Reference
Berries	Anthocyanins (68.6%), hydroxycinnamic acids (23.9%), flavonols (4.4%)	H_2O, membrane ultrafiltration	Preparative LC: 250 × 20 mm Eurospher 100-5 C_{18} Identification: was HPLC/DAD/ESI$^±$-MS^n, 150 × 2.1 mm, 5 µm. λ 280, 325, 360 and 520 nm	[98]
Costa Rican guava	Ellagic acid, myricetin, quercitrin, and quercetin	MeOH/H_2O (70 mL/100 mL). Freeze-dried pulp, mechanical dispersion	LC-TOF-ESI$^±$ (m/z range 100–1000), Synergi Hydro RP 80A 250 × 4.6 mm, 4 µm.	[110]
Brazilian guava, jambolan, nance, and lúcuma	Hydrolyzable and condensed tannins, flavonols, and flavanols	Acetone/H_2O/HCOOH (70:29:1). Freeze-dried pulp, accelerated solvent extraction	HPLC-DAD-ESI$^-$-MS^n, Aqua RP18 150 × 2.0 mm, 3 µm.	[113]
Perilla frutescents (L.) Britton	Rosmarinic acid (12.7–85.3%), scutellarein-7-O-glucuronide (6.5–45.1%), caffeic acid, apigenin-7-O-diglucuronide, and apigenin-7-O-glucuronide	Ethanol (EtOH)/H_2O (75 mL/100 mL). Accelerate solvent extraction (N_2 1200 psi 70 °C)	UPLC-PDA-ESI$^-$-TOF-MS, Kintex XB C_{18} column 150 × 2.1 mm, 1.7 µm	[114]
Solanum lycopersicum L.	e.g., caffeic acid hexosides, homovanillic acid hexoside, and dicaffeoylquinic acid (increasing trend)	Methanol (MeOH)/H_2O (80 mL/100 mL)	HPLC-DAD-ESI$^-$-MS/MS, Zorbax 300SB-C_{18} column (2.1 × 150 mm; 5 µm)	[115]
Rubus fruticosus L., *Prunus spinos* L. and *Cornus mas* L.	Gallic acid (138.0–443.5 mg kg^{-1} fresh weight), rutin (13.9–22.8 mg kg^{-1} fresh weight)	HCOOH/MeOH/H_2O (0.1/70/29.9)	LC-FLD λ$_{ex}$ 280, 320, 322 nm λ$_{em}$ 360 nm. Eclipse XDB C_{18} 150 × 4.6 mm	[116]
Green, herbal and fruit teas	Gallic acid, caffeic acid (+)-catechin, (−)-epicatechin, (−)-epigallocatechin, procyanidin B_1, and procyanidin B_2 contribute to 43.6–99.9%.	95 °C for 10 min	LC-PDA/FLD scan 260–400 nm absorbance matching Zorbax Eclipse XDB-C_{18}, 150 × 4.6 mm, 5 µm	[117]
Dried and candied fruit	Vanillic, ellagic, gallic, p-coumaric, chlorogenic, caffeic, ferulic, rosmarinic acids, and myricetin, quercetin, kaempferol, delphinidin, cyanidin, and pelargonidin	MeOH/H_2O (62.5 mL/100 mL). Sonication	HPLC-DAD at 260, 280, 329, and 520 nm. Zorbax Eclipse Plus C_{18} column 150 × 4.6 mm, 3.5 µm	[118]
Pink guava	Ellagitannins, flavones, flavonols, flavanols, proanthocyanidins, dihydrochalcones, and anthocyanidins, and non-flavonoids such as phenolic acid derivatives, stilbenes, acetophenones, and benzophenones	Freeze dried pulp, MeOH/H_2O (90:10), sonication	UHPLC-DAD-ESI$^+$-MS/MS, BHE Shield RP18 150 × 2.1 mm, 1.7 µm.	[119]
Blackberry juice		Microfiltrate (tubular ceramic membrane)	HPLC-DAD-ESI$^+$-IT-MS/MS Lichrosrb ODS-2 250 × 4.6 mm, 5 µm	[120]

Gordon and coworkers used accelerated solvent extraction (ASE) to characterize polyphenolic compounds in *Psidium guineense* Sw., *Syzygium cumini* (L.) Skeels, *Byrsomina crassifolia* (L.) Kunth, and *Pouteria macrophylla* (Lam.) Eyma. [113]. ASE techniques allow for multiple extractions simultaneously. Swifter assays are obtained which, in turn, expedite research results and minimize solvent waste [114] when compared to common extraction methods (e.g., Soxhlet, sonication). Anton and coworkers investigated the effect of ripening in tomato polyphenols content and antioxidant capability. A differential mass spectrometry approach allowed the authors to conclude that cultivar-dependent patterns are observed during ripening (e.g., maximum concentrations of polyphenols achieved half-ripe stage) [115]. Radovanović and coworkers, associated polyphenols from berries to antibacterial activity [116]. Veljković and coworkers analyzed phenolic compounds in different types of tea. Nettle/pineapple, and bearberry/raspberry teas showed the lowest and highest phenolic contents, respectively [117]. Miletić assessed polyphenols in dried and candied fruit. In this particular case, acid hydrolysis was applied to the previously dispersed methanolic extracts to free matrix-bound polyphenols [118]. One g *tert*-butyl hydroquinone/100 mL was added during extraction as a radical sink to protect polyphenols. Kowalska and coworkers used preparative chromatography to remove non-phenolics [98].

Tentative screening for *Psidium friedrichsthalianum* (Berg) Niedenzu pulp showed 1,5-dimethyl citrate, 1-*trans*-cinnamoyl-β-D-glucopyranoside, sinapic aldehyde-4-O-β-D-glucopyranoside, 1,3-O-diferuloylglycerol, and 3,3′,4-tri-O-methylellagic acid-4′-O-D-glucopyranoside [110]. Phenolic compounds from pink guava from Costa Rica have been recently reported, $n = 60$ phenolic compounds were characterized. The authors report for the first time in *P. guajava* $n = 42$ compounds in the fruit's peel and flesh, and $n = 24$ new compounds, e.g., phlorizin, nothofagin, astringin, chrysin-C-glucoside, valoneic acid bilactone, cinnamoyl-glucoside, and two dimethoxy cinnamoyl-hexosides [119]. During polyphenol analysis, HLB® SPE (Hydrophilic-Lipophilic,

Balance Solid Phase Extraction) cartridges are used routinely for clean-up. At least one research group has applied this approach to assay polyphenols and vitamin C in plant-derived materials [121]. Interestingly, when using the Folin–Ciocalteu spectrophotometric approach, ascorbate is considered interference and must be eliminated from the eluate (usually taking advantage of ascorbate thermolability) or else the measurements are overestimated. However, simultaneous retention of both analytes in the SPE cartridge can be exploited, if HPLC methods are used instead. We recommend that in countries in which fruits with high polyphenol content are readily available (and in considerable quantities), preparative separation of polyphenol fractions is a possibility for obtaining pure compounds (See for example, [122]). Finally, vanillic acid was reported in cocoa pod polyphenol-rich extracts. Interestingly, the application of 2000 mg L^{-1} of this cocoa extract to a vegetable oil improved its oxidative stability and shelf-life [123].

Method Application Experience

In our laboratory, ultrasound-assisted extraction is preferred for reducing processing time and avoiding degradation of the compounds. Additionally, polyphenols are quite light sensitive, hence yellow lights are used during the extraction using acetone-water or methanol-water solutions. As the polyphenol family is extensive and chemically diverse, a surface response design is always recommended to assess the appropriateness of the solvent system (i.e., selecting a solvent that provides the highest yields). Samples with a high lipid content (i.e., > 5 g total fat/100 g) usually cause significant interferences and must be defatted previous to polyphenol extraction. It is usual to add additional antioxidants (e.g., ascorbic acid) to polyphenol extracts to protect them from oxidation. Finally, it is common to find natural existing polyphenols as adducts with protein or carbohydrate moieties. These adducts are usually formed by non-covalent interactions (e.g., salt bridges); therefore, by adjusting the extract ionic strength, one can remove these artifacts. Sugar adducts are considerably more difficult to analyze since only a few compounds are commercially available (e.g., cyanidin 3-O-glucoside chloride). Hydrolysis (mild acidic, basic or enzymatic) is the usual approach to circumvent the lack of these commercial standards. Availability of mass spectrometry or nuclear magnetic resonance (NMR) can help elucidate unknown compounds and adducts.

2.2. Capsaicinoids

Capsaicinoids are plant metabolites from the *Capsicum* genus which give pungency to chili peppers [124]. Scoville scale which measures the spiciness of the fruits (originally, tested by sensory assays) is reported in function of capsaicin concentration (i.e., mg capsaicin kg^{-1} × 16 [125]). Today, the most reliable, rapid, and efficient method to identify and quantify capsaicinoids is HPLC. Measurement of this molecules is significant as a quality measure of chili pepper (22 domesticated varieties consumed regularly worldwide), a crop which is of significant cultural and global trade market value [126]. More than 20 different capsaicinoids have been described; the foremost capsaicinoids found in these plant structures include capsaicin and dihydrocapsaicin [127] (Figure 2).

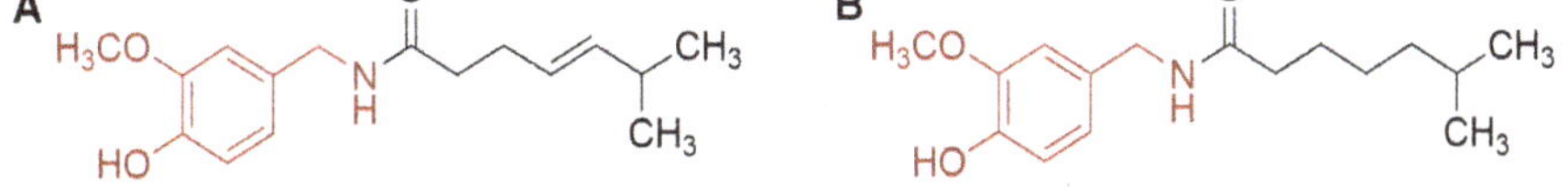

Figure 2. Chemical structures for (**A**) capsaicin (8-methyl-*N*-vanillylamide) and (**B**) dihydrocapsaicin (8-methyl-*N*-vanillylnonamide), the aromatic vanillyl radical is shown in red.

2.2.1. Measurement of Capsaicin and Dehydrocapsaicin in Real Samples

Research reports have described capsaicinoid analysis; the most recent are summarized in Table 5. Garcés-Claver and coworkers determined capsaicin and dihydrocapsaicin in two different scenarios, i.e., fruits grown in summer and then in spring [128]. The authors concluded that capsaicinoids varied largely among fruit families and that these families did not respond similarly to producing these capsaicinoids when their fruits were grown in the two seasons tested [128].

Table 5. Common chromatographic conditions used for capsaicinoid analysis.

Matrix	Extraction Method	Measurement Method, Chromatographic Column	Sensitivity, mg L^{-1} or mg kg^{-1} Fruit Dry Weight	Reference
Peppers *Capsicum annuum* L.	ACN, mechanical shaking	RP-LC/MS-TOF/ESI$^-$, pseudo-molecular ions [M-H]$^-$ 304.2 and 306.2 m/z. IS 4,5-dimethoxybenzyl)-4-methyloctamide, 250 × 4.6 mm, 5 µm	0.06	[128]
Natural capsaicinoid mixture (capsaicin/dihydrocapsaicin 67:33)	C_7H_{16}/EtOAc/MeOH/H_2O (1:1:1:1)	1. Sequential centrifugal partition chromatography. 2. Nucleosil 100-5 C_{18} column (125 × 3 mm, 5 µm, UV 280 nm	Preparative chemistry	[129]
Hot chilies, green peppers, red peppers, and yellow peppers	EtOH	HPLC-UV using a wavelength of 222 nm and a Betasil C_{18} 150 × 4.6 mm, 3 µm column	0.10	[130]
Vegetable and waste oils	Immunoaffinity column, SPE loading solvent, 5 mL MeOH/H_2O (5:95), washing solvent PBS, MeOH for elution	LC-ESI$^+$-MS/MS, Hypersil Gold, 100 × 2.1 mm, 3.0 µm	0.03	[131]
Edible and crude vegetable oils	SPE C_{18}, MeOH	IS capsaicin-d_3, and dihydrocapsaicin-d_3. RP-UPLC-ESI-MS/MS, ZORBAX Eclipse Plus C_{18} 50 × 2.1 mm, 1.8 µm)	0.5	[132]
Austrian chili peppers	ACN/H_2O (35:65)	UV and FLD λ_{ex} 280 and λ_{em} 310 nm, UPLC^{TM} BEH C_{18} 50 × 2.1 mm, 1.7 µm	0.136	[133]
Brazilian *Capsicum chinense* Jacq.	MeOH sonication	UHPLC–DAD–APCI-MS/MS, Hypersil Gold C_{18} 100 × 3 mm, 1.9 µm	0.0027	[134]
South Korean red peppers	MeOH/H_2O (95:5), 80 °C 2 h	FLD λ_{ex} 280 and λ_{em} 325 nm, Zorbax Eclipse XDB-C_{18} 75 × 3 mm, 3.5 µm)	0.06	[135]

SPE: Solid phase extraction. UPLC: Ultra-Performance Liquid-Chromatography.

Goll and coworkers optimized a cyclic solid support free liquid–liquid partition to separate a capsaicin and dehydrocapsaicin mixture into two sequentially collected product streams. This approach may serve as a base for compound purification before chemical characterization. With this optimization, the authors demonstrated theoretical and predictive tools are useful in preparative chemistry and process design [129].

The pretreatment of capsaicinoid determination (i.e., extraction steps) is usually straightforward, and the majority of methods are based on methanol-based extraction. However, Lu and coworkers reviewed several techniques that can be used to extract capsaicinoids successfully [136]. Ma and coworkers [131] used capsaicin and dihydrocapsaicin, and nonivamide [132] were selected as adulteration markers to authenticate vegetable oils. No capsaicinoid compounds were found in edible vegetable oils, thereby ruling out a possible adulteration source. The authors prepared immunosorbents by covalently coupling highly specific capsaicinoid polyclonal antibodies with CNBr-activated Sepharose 4B and packed into a polyethylene column [131]. This research is interesting, from the clean-up standpoint, since the authors adjusted the major parameters affecting the immunoaffinity column extraction efficiency (i.e., loading, washing, and eluting conditions) [131]. Schmidt and coworkers compared different chili peppers available in Austria and compared their contents of capsaicin and dihydrocapsaicin [133]. The authors used UPLC (Ultra-Performance Liquid-Chromatography) and hence obtained a reduced resolved chromatogram for both compounds of just 1.7 min. [133]. The authors also corroborated that the highest capsaicinoids content was in the fruits' placenta and the seeds. Similarly, Sganzerla and coworkers obtained a complete separation under 4 min [134]. The above examples correspond to high-throughput methods of analysis.

Finally, ingested capsaicinoids can persist in the bloodstream and can be determined in plasma using LC coupled with tandem mass spectrometry [137]. Intestinal absorption and metabolisms (via capsaicinoid glucuronides) have also been reported for a mammal [138]. At the same time, dietary capsaicin has been linked to the browning of adipose tissue, which in turn, promotes energy expenditure [139].

2.2.2. Method Application Experience

As shown, capsaicinoids can very well be measured by using a wavelength in the 200–400 nm UV range. However, fluorescence analysis can be performed (λ_{ex} 280 nm λ_{em} 338 nm) improving sensitivity dramatically [134], an approach preferred by our laboratory for routine analysis. A short column with a smaller particle size seems to improve both resolution and sensitivity.

2.3. *Caffeine and Theobromine*

Caffeine and theobromine are naturally occurring methylxanthines with antioxidant potential [140] (Figure 3). There are some misconceptions regarding health effects caused by caffeine ingestion [140]. On the contrary, theobromine (and cocoa) consumption has demonstrated beneficial effects [141]. Coffee, cocoa, tea, and caffeine-containing beverages (e.g., soft and energy drinks) are widespread and relevant food commodities. For example, caffeine intake has been calculated at 25 and 50 mg per day for children and adolescents aged 2–11 and 12–17 years, respectively. The more relevant caffeine sources were soda and tea as well as flavored dairy (for children aged < 12 years) and coffee (for those aged 12 years and above). Similarly, caffeine consumption has been between 2.5–3 and 400 mg kg^{-1} bw (body weight) day^{-1} for children and adults, respectively [142,143]. The evidence is suggesting an alimentary impact as some nutrients are poorly absorbed when combined with alkaloids [140]. Caffeine analysis is common in the food industry (e.g., quality control in beverages) and research (e.g., alkaloid carrying plants); it has also been incorporated in academia and student curricula [144].

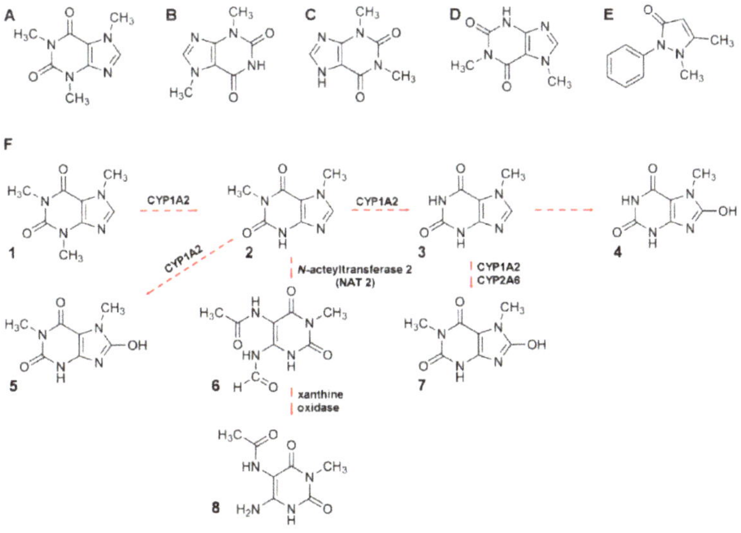

Figure 3. Chemical structures for (**A**) caffeine (1,3,7-trimethylxanthine), (**B**) theobromine (3,7-dimethylxanthine), (**C**) theophylline (1,3-dimethylxanthine), (**D**) paraxanthine (1,7-dimethylxanthine), and (**E**) antipyrine (2,3-Dimethyl-1-phenyl-3-pyrazoline-5-one or phenazone). (**F**) Caffeine biotransformation pathway is dependent on the CYP1A2 and CYP2A6 enzyme system. 1. 1,3,7-trimethylxanthine 2. 1,7-dimethylxanthine 3. 7-methylxanthine 4. 7-methyluric acid 5. 1-mthyluric acid 6. 5-acetylamino-6-formylamino-3-methyluracil 7. 1,7-dimethyluric acid 8. 5-acetylamino-6-amino-3-methyluracil [145].

2.3.1. Alkaloid Analysis and Reported Application to Real Samples

Several methods have been developed for alkaloid analysis in food samples. Also, methods for studying the fate of these alkaloids have been documented (Table 6). For example, Grujić-Letić and coworkers, analyzed 12 commercial tea and coffee products, non-alcoholic energy drinks and foods (including mate, green tea, and black tea), 5 combined preparations of over the counter non-steroid anti-inflammatories and water samples collected from 7 representative locations of the Danube River [146]. This paper represents a clear example of method versatility, as a single analyte was recovered, from variable matrices, and assessed using a similar procedure. This analysis was not only used for characterization, but also demonstrated a potential for quality control in commercial products (e.g., compliance of the nutritional label) and water. In water samples, the highest caffeine concentration found was 306.120 ± 0.082 ng L^{-1} during springtime. Gonçalves and coworkers recently demonstrated that caffeine might be a suitable chemical marker of domestic wastewater contamination in surface waters [147].

Table 6. Summary of conditions regarding alkaloid analysis.

Matrix	Extraction Method	Measurement Method, Chromatographic Column	Sensitivity, mg L^{-1} or mg kg^{-1}	Reference
Food Samples				
Energy drinks	Sonication for degassing	DAD 270 nm (caffeine) Nova-Pak C_{18} 150×3.9 mm, 5 µm, mobile phase: MeOH, NaH_2PO_4/hexanesulfonic acid ($C_6H_{13}SO_3H$)	0.023	[148]
Energy drinks	Sonication for degassing. "Dilute and shoot"	25 mmol L^{-1} NaAOc/HAOC buffer, pH 6.0, an inertsil OctaDecylSilane-3V 250×4.6 mm, 5 µm, UV 230 nm	0.19	[149]
Cocoa	Defat with C_6H_{14}, Acetone/H_2O/HAOc (70/29.5/0.5)	1. HPLC 250×4.60 mm, 5 µm 2. UPLC Acquity HSS T3 100×2.1 mm, 1.8 µm	0.001 for both LCs	[150]
Cocoa-based products	Defat by mechanical dispersion with C_6H_{14}, MeOH/H_2O (80:20)	UHPLC-Q-Orbitrap-MS/MS polyphenols ($n = 35$, ESI$^-$) and alkaloids ($n = 2$, ESI$^+$) Kinetex biphenyl 100×2.1 mm, 1.7 µm	Theobromine 0.03, caffeine 0.04	[151]
Mate beer and mate soft drinks	Sonication for degassing, ACN.	HP-TLC LiChrospher silica gel plates, fluorescence indicator and mobile phase acetone/toluene/chloroform (4:3:3) UV 274 nm	0.4	[152]
Biological Samples				
Human and synthetic plasma	Ultracentrifugation, 12,000 rpm	Waters Atlantis C_{18} 150×4.6 mm, 5 µm. Mobile phase: 15 mmol L^{-1} PBS (pH 3.5)/ACN (83:17). PDA 274 nm, IS: antipyrine	0.02	[153]
Human saliva	Chloroform/isopropanol (85:15)	Mobile phase: H_2O/HAOc/MeOH/ACN (79:1:20:2), Kromasil 100 C_{18} 250×4.6 mm, 5 µm, 30 °C, UV 273 nm	0.032	[154]
Human and neonate plasma	SPE polymeric 96-well plates Strata-X™. Elution: MeOH/H_2O/HAOc (70:29:1)	10 mmol L^{-1} PBS (pH 6.8)/ACN (gradient mode). Zorbax® SB-Aq narrow bore RR 100×2.1 mm, 3.5 µm), 40 °C, UV 273 nm	0.1	[155]

Shrestha and coworkers developed a method for use as quality control. Concentrations of Nepalese tea and coffee ranged from 1.10 to 4.30 mg caffeine kg^{-1} dry basis [156]. Fajara and Susanti also determined caffeine in coffee beverages; they found 109.7–147.7 mg caffeine kg^{-1} per serving [157]. Gliszczyńska-Świgło and Rybicka used both a photodiode and fluorescence detector to monitor both caffeine and water-soluble vitamins, simultaneously, in energy drinks [148]. Aşçı and coworkers analyzed caffeine in soft drinks [158]. The authors used Behnken response surface design to optimize HPLC conditions. Optimized variables included pH, 6.0, flow rate, 1.0 mL min^{-1} and a mobile phase ratio, 95% [158]. Similarly, preservatives sorbate and benzoate also can be determined with caffeine simultaneously in sports drinks [149]. Ortega and coworkers compared data from HPLC- and UPLC-MS/MS (MS/MS also known as tandem mass spectrometry). The authors analyzed procyanidin oligomers (mono to nonamers) and catechin, epicatechin, caffeine, theobromine. The analysis was performed under 12.5 min [150]. Recently, Rodríguez-Carrasco and coworkers used to analyze polyphenols and alkaloids in cocoa-based products. Mainly, they compared three different coffee varieties including "Forastero", "Trinitario", and "Criollo". Mostly, theobromine was found in major quantities relative to caffeine except Criollo 70 and 75% where the theobromine/caffeine ratio is ca. 1:1. Of all samples examined, Criollo varieties showed the highest quantities of alkaloids. [151]. Interestingly, a positive association has been described between cacao polyphenol absorption and theobromine [159]. Other identifying markers, such as fatty acids, have also been reported as tools for discrimination among coffee varieties. The authors were able to discern *Coffea arabica* (Arabica) and *Coffea canephora* (Robusta) using $\sum$MUFA, 18:3n3, $\sum$MUFA/$\sum$SFA [160].

2.3.2. Alkaloid Bioavailability and Transference to Biological Samples

Caffeine is rapidly absorbed following oral consumption; maximum blood (plasma) levels are usually reached within 30 min [140]. Caffeine bioavailability studies have been performed in human plasma, for example, Alvi and Hummami monitored caffeine and antipyrine (Figure 3). Caffeine in human plasma was stable for at least 24 h at room temperature or 12 weeks at $-20\,^\circ$C [153]. Caffeine is a demonstrated therapeutic agent for apnea of prematurity. Hence, López-Sánchez developed a method to monitor caffeine in serum to demonstrate that the drug had achieved its therapeutic levels (i.e., 30 or 35 µg mL^{-1}) [161]. Cleanup using SPE adapted in multiple well plates, as the one used in the former study, is an easy way to process several samples simultaneously, instead of the one-on-one cartridge approach. Only in 85% and 78% of the cases studied, maternal and newborn absorption of caffeine was demonstrated, respectively. Another research group investigated caffeine metabolism based on CYP1A2 enzyme activity. The presence and ratio of theophylline, paraxanthine, theobromine, and caffeine (Figure 3) was evaluated in human saliva [154]. The authors collected saliva of healthy subjects after consumption of a caffeinated beverage and obtained data of compared chromatographic profiles from the saliva of smoking (active xenobiotic hepatic metabolism) and non-smoking subjects [154]. Saliva, plasma, and urine already have been demonstrated valuable to intervention studies for cocoa [155,162]. Kobayashi used differential chromatogram analysis to narrow the signal width for caffeine, in urine samples, to improve separation demonstrating that peak enhancing posterior to injection is possible [163]. Finally, Ramdani and co-workers incorporated green and black tea powder into bovine diets demonstrating that alkaloids, catechins, and theaflavins diminished ammonia and methane productions without any detrimental effect on rumen functions *in vitro* [51].

Although theobromine is not a usual analyte for feed analysis, is noteworthy that the 2002/EC/32 regulation sets limits for the analyte at 300 mg kg^{-1} for compound feed, except for adult cattle feed, where the threshold is laxer (i.e., 700 mg kg^{-1}).

2.3.3. Method Application Experience

Tea and coffee sample extracts are rich in tannins and other non-desired compounds that may generate matrix effects and reduce the shelf life of an analytical column. We have successfully used MgO to remove said interferences while increasing the extract pH. An alkaline medium ensures positively charged alkaloid molecules. Furthermore, defatting is vital for an adequate recovery when a lipid-rich sample is treated (e.g., cacao seeds, >30 g total fat/100 g), especially, if aqueous extracting is employed. We suggest the use of efficient organic solvents; *n*-hexane, petroleum benzine, for example, have been exploited. Minimal amounts possible should be used, as this otherwise generates waste. Chlorinated solvents and ethyl ether should be avoided, as alkaloids exhibit some degree of solubility in these solvents which, in turn, may affect recovery.

2.4. Cholesterol

Cholesterol ((3*S*,8*S*,9*S*,10*R*,13*R*,14*S*,17*R*)-10,13-dimethyl-17-[(2*R*)-6-methylheptan-2-yl]2,3,4,7,8, 9,11,12,14,15,16,17-dodecahydro-1*H*-cyclopenta[a]phenanthren-3-ol), is a waxy steroid metabolite found in the cell membranes and transported in the blood plasma of all animals [164]. This sterol plays a role in metabolic (e.g., precursor for bile acids and steroid hormones) and structural processes (e.g., regulates biological membrane fluidity) [165,166]. Cholesterol can be introduced to the metabolism through *de novo* synthesis or diet [162]. In plant structures, similar compounds are found such as phytosterols and stanols [167]. However, when analyzing cholesterol, one must consider that the amount of cholesterol made by many plants is not negligible [168]. Nutritional information regarding cholesterol content in food and intake through dietary sources is relevant, as overload can drastically increase plasma cholesterol levels and, hence, health risks. From a methodological standpoint, a considerable advantage in using the LC approach is that lipid oxidation is negligible, as measurements can be performed at relatively low temperatures. Herein are detailed some examples of cholesterol analysis in food samples (Table 7).

Albuquerque and coworkers compared both HPLC and UPLC for the analysis of eggs, egg yolks, sour cream, and chicken nuggets. The latter approach rendered a method with 8-fold less solvent waste and ca. 4-fold more sensitivity, with a decreased analysis time (i.e., 4 min) [166]. The initial sample mass used from the assay was optimized; 0.25 g and 1 g for samples with relative lower (e.g., sour cream) and higher (e.g., egg yolk) cholesterol contents. The authors also compared different cooking methods for the chicken nuggets (baked vs. deep frying). They found that cholesterol content was higher in the oven baked goods. This is a result of the processing as the meat loses water during baking. Meanwhile, water/oil exchange occurs during frying. Although several solvents were tested, the authors concluded that an acetonitrile/2-propanol solvent system was the most successful in eluting the cholesterol molecule [166]. Cholesterol analysis usually renders clean chromatograms since most interferences are eliminated by saponification. Saponification segregates the molecule of interest from the saponifiable lipid fraction (e.g., acylglycerols) and hydrolyzes cholesterol esters. This step has been considered critical for cholesterol analysis in food matrices [166]. Furthermore, Cruz and coworkers, quantitatively, compared several extraction methods on freeze dried and thawed seafood samples [169]. In this regard, the direct saponification and extraction considerably reduce solvent waste, while the Smedes method used non-chlorinated solvents (is a greener approach). Better recoveries for vitamin E are obtained when the analysis is performed before saponification step (e.g., modified Folch, Smedes). The authors were able to analyze α-tocopherol, cholesterol, and fatty acids all from the same extract and applied the optimized method to octopus, squid, mackerel, and sardine successfully. From the assayed samples, squid and sardine showed higher values of cholesterol and vitamin E, respectively. Interestingly, normal phase chromatography was used to assess vitamin E [169]. Saldanha and Bragagnolo also used normal phase chromatography. The authors used very mild conditions during saponification, which are paramount to avoid cholesterol oxidation. Also, they monitored cholesterol contents after heat treatment and demonstrated that it decreased significantly, with a simultaneous increase of the cholesterol oxides contents (i.e., 19-hydroxycholesterol, 24(*S*)-hydroxycholesterol, 22(*S*)-hydroxycholesterol, 25-hydroxycholesterol, 25(*R*)-hydroxycholesterol, and 7-ketocholesterol) [170]. Bauer and coworkers analyzed cholesterol and cholesterol oxides in milk samples using reversed-phase chromatography. [171]. The presence of cholesterol oxides can indicate the source and nature of the food, as well as the storage and processing conditions suffered by a commodity. The authors conclude that milk has physicochemical characteristics that make it more resistant to oxidation of cholesterol compared to other products of animal origin. In this regard, several sample preparation methods for cholesterol oxides have been detailed elsewhere [173]. Daneshfar and coworkers used dispersive liquid–liquid microextraction as an alternative to the extraction and clean-up steps in sample preparation [172]. In this case, ethanol was used as a disperser solvent and carbon tetrachloride as an extraction solvent [172]. This work is a fine example of parameter optimization during method validation; different dispersion (i.e., EtOH, acetone, and ACN) and extraction (i.e., CS_2, CH_2Cl_2, $CHCl_3$, and CCl_4) solvents were tested, as well as variables such as pH, volume and time. However, the authors fail to explain how they obtain total cholesterol from a complex matrix (for example, a method must be able to free cholesterol from its esterified form) when no hydrolysis is performed (i.e., ensuring not just the mere quantification of unbound/free cholesterol).

Table 7. Measurement techniques meant for cholesterol in food samples.

Matrix	Extraction Method	Measurement Method, Chromatographic Column	Sensitivity, mg L^{-1} or mg kg^{-1}	Reference
Egg-, dairy- and meat-based products	ACN/2-propanol	1. Supelcosil™ LC-18-DB 150 × 4.6 mm, 3 μm	3	[166]
		2. Acquity UPLC® BEH C_{18} 50 × 2.1 mm, 1.7 μm, UV 210 nm	0.7	
Seafood	1. In situ: KOH 2 mol L^{-1}/MeOH, 80 °C, N_2, C_6H_{14} 2. Modified Folch: MeOH/CH_2Cl_2 (1:2), saponification 3. Smedes: 2-propanol/cyclohexane (1:1.25), saponification	Vitamin E: FLD λ_{ex} 290 λ_{em} 330). Cholesterol: UV 210 nm, Supelcosil™ LCSI 75 × 3.0 mm, 3 μm, mobile phase: n-hexane and 1,4-dioxane (97.5:2.5) IS: tocol	Vitamin E: 0.05 Cholesterol: 10	[169]
Seafood	KOH 50 g/100 g/EtOH, 25 °C, 22 h, in the dark, C_6H_{14}	1. Nova Pack CN HP 300 × 3.9 mm, 4 μm, n-hexane/2-propanol (97:3), UV 210 nm (cholesterol oxides), RID (cholesterol and epoxides) 2. Confirmation: HPLC-APCI-MS QTRAP®	0.01	[170]
Dairy product	KOH 50 g/100 g/EtOH, 25 °C, 22 h, in the dark, C_6H_{14}	Restek C_{18} 150 × 6 mm, 5μm, mobile phase: ACN/2-propanol (95:5), UV 202 nm 25-hydroxy and cholesterol, 227 nm 7-ketocholesterol	11.10	[171]
Egg and dairy product and vegetable oil	1. Egg yolk and milk: pretreatment with ACN 2. Liquid–liquid dispersion (DLLME) EtOH (800 μL)/CCl_4 (35 μL).	CLC-ODS-C_8 150 × 6 mm, 5μm. Mobile phase: ACN/EtOH (50:50), HPLC-UV 210 nm,	0.01	[172]

It should be pointed out that though the chlorinated solvents are used in very small quantities, they are still classified by the IARC (International Agency for Research on Cancer) as possible human carcinogens (group 2B). Finally, Robinet and coworkers used a cholesterol esterase in an unrelated matrix to avoid chemical saponification [174]. In this regard, cholesterol esterases (most active at pH 7.0, 37 °C, and in the presence of taurocholate) and lipases (most active at pH 7.7, and 37 °C [175]) are commercially available.

Method Application Experience

We suggest two major points: i. that it is recommendable to perform the saponification first and then the solvent-aided extraction ii. a response surface design may be useful to optimize the length of the saponification treatment.

3. Determinations Designed for Feed and Feed Ingredients

3.1. Mycotoxins

3.1.1. Recent Approaches for the Determination of Mycoxotins in Feeds

Mycotoxins are secondary metabolites mainly by fungi *Aspergillus, Penicillium, Fusarium* and *Alternaria* species, in stress situations, which involve changes in temperature, moisture or pH in plants [58,176,177]. Currently there are more than 400 types of mycotoxins as ubiquitous contaminants in a wide variety of foods [178,179], such as, corn, cocoa, sorghum, wheat, oats, rye, cotton, peanuts, coffee, dairy products, eggs, among others [180]. Among the best known are ochratoxin (OTA), zearalenone (ZEA), trichothecenes, aflatoxin B_1 (AFB$_1$), fumonisin B_1 (FB$_1$) and their metabolites. The last two are listed as carcinogenic by the IARC [181]. Mycotoxins, in general, are teratogenic, mutagenic, carcinogenic, and can possess an immunosuppressive effect in both animals and humans [178,182], which can be aggravated by factors such as the animal species, the concentration of the toxin and synergism existing among them, in addition to the health and nutritional status of the animal [182,183]. Also, the direct effects on health, including decreased weight gain, feed conversion inefficiency, reduced production, and a decrease of the food system profitability, the increase in feedstuff costs, medical treatments, and ineffectiveness when exploiting the genetic potential of animals [183].

At an organ level, in the liver, AFB_1 can generate several metabolites, which include aflatoxin M_1 (AFM_1), which is transferred to milk, a complete food nutritionally, and which is vital in the development of the first years of life [184,185]. Also, the AFM_1 is a compound declared as a carcinogen that is very resistant to pasteurization and freezing [180,183]. Therefore, being trawl compounds in the trophic chain, which involve the adverse effects on livestock production, with an obvious risk to the health of consumers, it stresses the need for laboratories to possess the ability to analyze a large number of analytes in a single sample. In this way, the amount of information can be increased, and a wider diagnosis can be made about the safety of the food and feed industry.

In this regard, Table 8 shows a summary of methods developed for the identification and quantification of mycotoxins, by different research groups, focused mainly on animal feed. For example, Njumbe Ediage and coworkers developed a technique capable of determining 25 mycotoxins in cassava meal, peanut cakes, cornmeal, and different sorghum varieties. The most exciting thing, in this case, is how the researchers solved the affinity fact of fumonisin and ochratoxin with the amino groups (due to the presence of carboxylic acid moiety, Figure 4) [177,186]. The researcher divided their extract into two portions, one to which formic acid and dichloromethane were added. After cleanup, the two independent shares were remixed evaporated at 40 °C, reconstituted with $MeOH/H_2O/CH_3COOH$, and 5 mmol L^{-1} CH_3COO^- NH_4^+. During MS-based mycotoxin separations, flows are usually kept low, so solvent nebulization and evaporation are performed swiftly. The mobile phase is generally accompanied by an acetic or formic acid buffer to improve ionization especially for those compounds without readily ionizable functional groups (e.g., aflatoxins). Also, the formate ion is added in both solvents as one solvent depletes during the gradient separation and the buffer must always be present in a similar proportion [177,186]. Dzuman and coworkers and Rasmussen and coworkers, used, as an extraction method, a modification of the QuEChERS method, (Quick, Easy, Cheap, Effective Rugged, and Safe usually used for pesticide analysis). Both research groups coincide that QuEChERS adaptations for mycotoxin analysis open the possibility toward the simultaneous assay of several and distinct groups of contaminants (e.g., pesticides and mycotoxins) [179,187].

Table 8. Measurement techniques meant for mycotoxins in feed samples.

Matrix	Number of Analytes/Execution Time (min)	Extraction Method	Measurement Method, Chromatographic Column	Reference
Cassava meal, peanut cakes, cornmeal, and different sorghum varieties	25/28	$MeOH/CH_3CO_2CH_2CH_3/H_2O$ (70:20:10), cleanup was performed using amino SPE cartridges	LC: Symmetry RP-18 150 × 2.1 mm, 5 μm, Identification: MS/MS/ESI$^+$	[177,186]
Cereals, compound feed and silages	56/50	Modified QuEChERS method	LC: Acquity UP3 HSS T3 100 × 2.1 mm, 1.8 μm, Identification: MS/MS/ESI$^\pm$	[179]
Bovine milk	10/30	Acid acidified ACN and sodium acetate was used to separate the aqueous from the hydrophilic phase from milk	LC: Ascentis Express C_{18}, 150 × 2.1 mm, 2.7 μm, Identification: MS/MS/ESI$^+$	[185]
Silage	27/44	Modified QuEChERS method	LC: Gemini® C_6-Phenyl 100 × 2.0 mm, 3 μm, Identification: MS/MS/ESI$^\pm$	[187]
Millet and Sorghum	84 and 62 respectively/Not Indicated	$ACN/H_2O/CH_3COOH$ (79:20:1) mixture	LC: Gemini® C_{18}, 150 × 4.6 mm, 5 μm, Identification: MS/MS/ESI$^\pm$	[188,189]

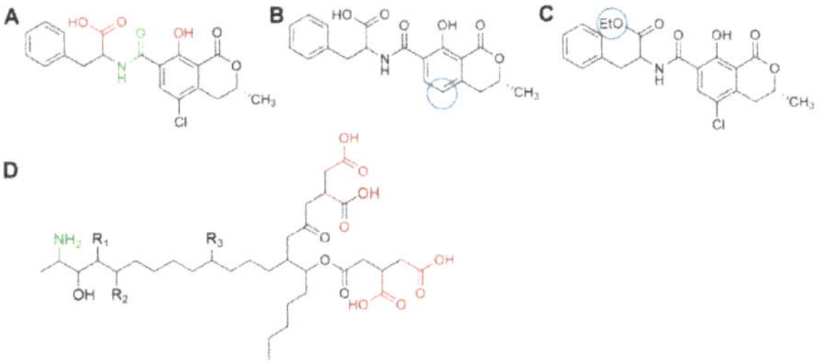

Figure 4. Chemical structures for (**A**) ochratoxin A, (**B**) ochratoxin B, (**C**) ochratoxin C, blue colored circles represent changes in the structure between ochratoxins, loss of Cl and OH in ochratoxin B and C respectively render a more lipophilic molecule. Et = C_2H_5, and (**D**) are the general backbone of Fumonisins. FB_1 = 721.83 g mol^{-1} R_1: H R_2: OH R_3: OH; FB_2 = 705.84 g mol^{-1} R_1: OH R_2: H R_3: OH; FB_3 = 705.84 g mol^{-1} R_1: H R_2: H R_3: OH; FB_4 = 689.84 g mol^{-1} R_1: H R_2: H R_3: H. Functional groups colored in green and red represent a positively and negatively ionizable moiety, respectively.

3.1.2. Agricultural by-Products as Feed Ingredients

Agricultural and food-industry residues are valuable to animal nutrition as they are rich in many bioactive and nutraceutical compounds, such as polyphenolics, carotenoids and dietary fiber among others [190]. Agro-byproducts, used in animal feed, originate from perishable crops and, as such, are susceptible to fungal infection [191]. Hence, mycotoxin surveillance of these materials contemplating the most common contaminants present in such matrices, but also considering emerging contaminants (e.g., beauvericin, enniantins, and fusaproliferin) [191,192] is paramount. The food industry generally includes practices that guarantee the safety of the product meant for human consumption. Residues destined for animal production may not be subject to the same scrutiny. For example, the wine industry with a production estimated at 27 million liters worldwide. Presence of OTA in wine has been widely investigated [193]. However, with the development of new methods, it has been possible to find up to 36 different mycotoxins. (See for example, [194]).

Countries where the production of wine is the predominant, compared to other types of industry, a considerable amount of waste must be repurposed. As such, this might be of use as a ruminant (such as cows and goats) feed ingredient, where the pulp, husks, and seeds, might offer to the animal diet: fiber, energy, fatty acids, and antioxidant compounds which improve ruminal health, echoing in the quality of meat and milk [195–197]. As yet another benefit from this waste processing, the use of grape seeds as mycotoxin adsorbents has been investigated both *in vitro* [198] and *in vivo* (e.g., pigs [199]).

3.2. Antibiotics

3.2.1. Recent Multiresidue and Multi-Class Analysis of Antibiotics in Feeds

Antibiotics are bioactive substances used against bacteria as a therapeutic, metaphylaxis or prophylactic agent both in humans and animals [200–202]. In livestock, some antibiotics are included in animal diets as growth promoter (e.g., monensin, narasin, ractopamine), decrease feed conversion, improve feed efficiency, and overall cost-effectiveness of animal production systems [203,204]. Overuse of veterinary pharmaceuticals in livestock, aquaculture, and the feed industry is reflected in the incidence of residues found in animal-derived food products (e.g., meat, eggs, milk, and honey) [201,205–207]. Antibiotic biotransference through the food chain may contribute to allergic reactions, mutanogenic and cancerogenic effects, found in humans and animals; additional to the growing rates of antimicrobial resistance [208,209]. Considering these issues, organizations worldwide (e.g., European Commission, United States Food and Drug Administration, World Health Organization) have generated protocols that help control, regulate and surveil the use of antibiotics in food-producing animals [208,210–212]. Hence, similar to mycotoxins, development of analytic methods that allow for identifying and quantitating a broad spectrum of compounds from a sample, directly contributes to surveillance programs for feedstuff manufacturing (raw materials or feed ingredients, compound feed, and premixes) and, similarly, those commodities derived from food-producing animals.

Table 9 shows a summary of the different characteristics of validated methods for the identification and quantification of veterinary antibiotics in different types of matrices. Molognoni and coworkers, optimized a method for the determination of spectinomycin, halquinol y zilpaterol in compound feed demonstrating once again the capabilities of mass spectrometry to assess two or more families of seemingly unrelated compounds. The authors tried both hydrophilic interaction and reverse-phase chromatography. Though HILIC (Hydrophilic Interaction Liquid Chromatography) offered good results, it requires a longer analysis time (i.e., up to 5 additional min), and is pH sensitive. Reverse-phase chromatography requires a relatively inexpensive column that is usually available in laboratories and which analytical instrumentation providers generally keep in stock. Additionally, a more effective separation was archived using heptafluorobutyric acid in the mobile phase [202].

Table 9. Measurement techniques meant for veterinary antibiotics in food and feed samples.

Matrix	Number of Analytes/Execution Time	Extraction Method	Measurement Method, Chromatographic Column	Reference
Recent Multiresidue and Multi-Class Analysis of Antibiotics in Feeds				
Rendering products	40/Not Indicated	During extraction, fat was removed and clean up performed using an SPE PRiME HLB cartridge, eluate evaporated to dryness and reconstituted with ACN and formic acid	BEH C_{18} column Identification: HPLC-MS/MS/ESI$^+$	[201]
Compound feed	3/Not Indicated	Formic acid/H_2O (80:10)	Hypersil Gold HILIC (150 × 3.0 mm, 5 μm) and C_{18} (50 × 2.1 mm, 3.5 μm). Identification: HPLC-MS/MS/ESI$^+$	[202]
Pig, poultry, and cattle feed	62/13	ACN/H_2O (90:10) acidified with CH_3COOH.	C_{18} Vensusil XBP (50 × 2.1 mm, 3.0 μm, 100 Å). Identification: HPLC-MS/MS/ESI$^+$	[208]
Feed	10/Not Indicated	Acidic extraction with hydrochloric acid (0.5 mol L^{-1} aqueous solution), and purified by SPE cartridge	Acquity UPLC HSS T3 (150 × 2.1 mm, 1.7 μm). Identification: HPLC-MS/MS/ESI$^\pm$	[213]
Multiresidue Analysis of Antibiotics in Foods				
Fish muscle	41/20	Extraction with ammonium formate and ACN/H_2O (80:20)	X-SELECT C_{18} (150 × 2.1 mm, 3.5 μm) Identification: HPLC-MS/MS/ESI$^\pm$	[205]
Shrimp	24/8	Extraction with formic acid in water and ACN	XBridge BEH C_{18} (100 × 2.1 mm, 2.5 μm). Identification HPLC-MS/MS/ESI$^+$	[206]
Poultry muscle tissue and eggs	14/14	ACN extraction Centrifugation at 0 °C 45 min	Poroshell 120 ECC$_{18}$ (50 × 3.0 mm, 2.7 μm) Identification: HPLC-MS/MS/ESI$^\pm$ (quadrupole linear ion trap)	[207]
Honey	6/Not Indicated	Modified QuEChERS method Extraction was performed using ACN and $MgSO_4$ and NaCl	ZORBAX Eclipse XDB C-18 (150 × 4.6 mm, 5 μm). Identification: HPLC-MS/MS/ESI$^+$	[214]

ACN: Acetonitrile.

3.2.2. Multiresidue Analysis of Antibiotics in Foods

Barreto and coworkers developed a method to assay $n = 14$ different coccidiostats (i.e., lasalocid A, maduramicin, monensin, narasin, salinomycin, semduramicin, robenidine, diclazuril, toltrazuril, trimethoprim, chlopidol, amprolium, diaveridin y nicarbazin) in poultry muscle tissue and eggs; after testing several chromatographic columns, they selected the one that completed the separation under less time (i.e., 14 min). The authors used low temperature clean-up as an alternative to SPE, reducing costs, time and ion suppression. Internal standards where used to compensate intense matrix effects [207]. Regarding aquaculture, Kang and coworkers analyzed $n = 41$ antibiotics in fish muscle [205]. Similarly, Kumar Saxena and coworkers developed and validated $n = 24$ antibiotics (including quinolones, sulfonamides, and tetracyclines) in shrimp, and they preferred to use methanolic separation [206]. Finally, Shendy and coworkers identified $n = 6$ different classes of antibiotics in honey with a modified QuEChERs procedure simultaneously. Extraction was performed using ACN and $MgSO_4$ and NaCl [214].

For both mycotoxins and antibiotics, a review was made of the wide variety of methods used in the food industry for the simultaneous, extraction of multiple analytes. For the identification and quantification of each chemical, a sensitive and selective tool is required. It is here that mass spectrometry has been useful, by reducing costs and response time. [185,202,209].

3.2.3. Method Application Experience (Mycotoxins and Antibiotics)

A multitoxin (n = 26) analysis was applied to feedingstuffs using, as a reference, a method previously described by Wang and coworkers in cornmeal. $ACN/CH_3COOH/H_2O$ (74:1:25) was used for extraction and cleanup we exploited the versatility of HLB cartridges (which allow the retention of a wide array of analytes with the least of interferences) [215]. When compared with immunoaffinity columns, this sorbent is less prone to fracturing and do not require low temperatures for storage. Later, the recovered extract was evaporated to dryness using vacuum at 60 °C and reconstituted with MeOH. The method relies on the 12.5-fold concentration of the original analyte to improve sensitivity. In the case of antibiotics (n = 23), we based our procedure on that described by Duelge and coworkers [216]. We extracted and eluted analytes using an ACN/MeOH solution. Again, we trusted the versatility of HLB SPE cartridges during cleanup. Both assays were single quadrupole equipped LC system using ESI^+ and relied on a reverse phase separation (Zorbax Eclipse Plus, 100 × 3 mm, 3.5 μm). For mycotoxin separation, the mobile phase consisted of a gradient using acidified (0.1 mL/100 mL formic acid) ACN and H_2O. For antibiotics, the gradient consisted of three different acidifed solvents ACN, H_2O, and MeOH. In our experience, the first two-solvent gradient (starting with water) can separate most antibiotic families (β-lactams, tetracyclines, macrolides, streptogramins, lincosamides, aminoglycosides). Our gradient finishes with MeOH which is the only solvent capable of eluting coccidiostats (e.g., monensin and narasin). Efficient chromatographic separation was achieved under 35 min.

3.3. Amino Acids

Protein building blocks (i.e., amino acids), biologically, can be separated into two main groups. Exogenic/essential amino acids (i.e., Arg, Phe, His, Ile, Leu, Lys, Met, Thr, Trp, and Val), are not synthesized by the organism and must be provided in the diet to cover the requirement. The remaining amino acids are endogenic (i.e., Ala, Cys, Asp, Glu, Pro, Ser, Tyr, and Gly). Several of these amino acids (e.g., Lys, Met, Thr, and Trp) are prepared synthetically and are commercially available to use as feed additives. The purity of these additives must be routinely checked and adequately verified. Hence, methodological development is paramount for quality control for determination of amino acids in feed materials and feed mixtures. However, few reports have focused on feed. As a result; we intend to give an overview of the methods available in related matrixes.

3.3.1. Fish Tissue

In a comprehensive research article, Mohanty and coworkers reported the complete amino acid profile (except tryptophan which was assessed spectrophotometrically and basic hydrolysis) for 27 different food fishes. [217]. Derivatization was performed using 6-aminoquinolyl-N-hydroxysucciminidyl carbamate (AQC), this specific reagent requires neutral pH to work. Adduct formation has the advantage of being stable and reacting with secondary amines. No variability among profiles was found in fishes of the same species from different locations. They also related the concentration of the amino acid found in the fish with the environment in which they live (e.g., marine and cold-water fishes showed relatively higher amounts of Met). At the same time, they recommend the consumption of certain fish species for several amino acids dietary deficiency in humans [217]. Example of methods suitable to analyze amino acids in diverse matrixes is shown in Table 10.

Table 10. Sample pretreatment, derivatization and measurement conditions for amino acids in feeds.

Matrix	Hydrolysis	Derivatization	Measurement Method, Chromatographic Column	Reference
		Applications in Feed and Related Matrices		
Spirulina sp.	Various physical methods	2-mercaptoethanol	Licrospher 100 RP 18 125 × 4 mm, FLD λ_{ex} 360 λ_{em} 460 nm	[218]
Spirulina sp.	1. Total AA: $HClO_4$ 8 mol L^{-1}, 150 °C for 2 h, 140 °C for 4 h, 120 °C for 8 h, and 110 °C for 22 h. 2. Free AA: CCl_3COOH, sodium deoxycholate	Triethylendiamine (TEA), phenylisothiocianate (separation of protonated species)	Supelcosil LC_{18}-DB 250 × 4.6 mm, 5 µm. Gradient 0.7 mol L^{-1} acetate buffer pH 6.4/TEA, H_2O and ACN/H_2O (80:20). UV λ 254 nm	[219]
Spirulina sp.	Pyrogallol, HCl 8.3 mol L^{-1} 70–80 °C 2 h, IS triundecanoin	o-phtaldialdehyde (OPA)	Zorbax AAA at 40 °C 40 mmol L^{-1} NaH_2PO_4 pH 7.8, $ACN/MeOH/H_2O$ (45:45:10), 2.0 mL min^{-1}, UV λ 338 nm	[220]
Plants	Soncation, EZ:faast™ Free Amino Acid Kit	propyl chloroformate	UHPLC EZ:faast™ 4u AAA-MS, 250 × 2.0 mm, 3 µm. IS: homoarginine, methionine-d_3, and homophenylalanine	[221]
Chamomile flowers	Free amino acids: Sonication	AccQ Fluor, 55 °C	Shimpack column (250 × 4.6 mm, 5 µm) FLD λ_{ex} 250 λ_{em} 395 nm	[222]
Rapeseed meal	HCl 6 mol L^{-1}, 110 °C 23 h	Ninhydrin	Ion exchange chromatography, Vis 570 nm (Pro 440 nm), IS: Norleucine	[223]
Feed ingredients	HCl 6 mol L^{-1} 0.1 g/100 mL phenol, 150 °C 6 h, Reacti-Therm™	Borate buffer pH 10, OPA (primary-) and FMOC (secondary amines)	Zorbax Eclipse-AAA 40 °C, λ_{ex} 262 λ_{em} 338 nm	[224]
Feed	6 mol L^{-1} HCl 110 °C 16–23 h, peformic acid, HBr,	AQC, borate buffer	AccQ-Tag Ultra C-18 100 × 2.1 mm, 1.7 µm). UPLC PDA 260 nm, IS: DL-2-aminobutyric acid	[225]
Fish tissue	6 mol L^{-1} HCl 110 °C closed vessel 24 h	AccQ-Fluor Reagent (AQC in 0.2 mol L^{-1} borate buffer pH 8.8)	FLD λ_{ex} 250 λ_{em} 395 nm. RP C_{18}	[217]
		Selected Applications		
Matrix	Hydrolysis and treatment		Measurement method, chromatographic column	Reference
Lipoprotein	Sodium dodecyl sulfate (SDS), enzymatic digestion (e.g., pronase E, muramidase)		UPLC BEH C_{18} 50 × 2.1 mm, 1.7 µm, 130 Å, UV 202–208 nm. Phosphate buffer 50 mmol L^{-1} pH 4.35/sodium azide and Phosphate buffer 75 mmol L^{-1} pH 4.95/MeOH (85:15)	[226]
Peptidoglycan	SDS, sonication, DNAse, RNAse, and trypsin. HCl for teichoic acids. Hydrolases (mutanolysin)		$CF_3COOH/MeOH$, UPLC-TOF/MS-ESI^+ CSH C_{18} 100 × 2.1 mm, 1.7 µm, UV 210 nm	[227]
Cocoa beans	Fermentation, HCl 0.1 mol L^{-1}, mechanical dispersion, ethyl ether		UPLC-ESI^+-MS Acquity UPLC BEH C_{18} 150 × 2.1 mm, 1.7 µm and LC/ESI^+-MS/MS Aeris Peptide XB-C_{18} 150 × 2.1 mm, 3µm	[228]
Olive seeds	n-hexane defat, Tris/HCl pH 7.5, SDS, dithiothreitol, high-intensity focus ultrasound, acetone precipitation, alcalase hydrolysis		RP-HPLC Jupiter Proteo 250 × 10 mm, 4 µm FLD λ_{ex} 280 λ_{em} = 360 nm RP-HPLC-ESI^+-QTOF-MS, Ascentis Express Peptide ES-C_{18} 100 × 2.1 mm, 2.7 µm	[229]

3.3.2. Filamentous Cyanobacteria, *Spirulina* sp.

Spirulina sp. is a filamentous cyanobacterium that have been recognized for its nutritional value as a feed ingredient and supplement, and has been related to health benefits in humans [230]. Its nutrient profile has been reported previously, and it has even exhibited a higher amino acid value (except for Lys, Glu, Pro, His) when compared with that of soybean meal (a staple feed ingredient) [218]. Additionally, based on this profile, they calculated energy for a broiler diet. Nurchaya Dewi and coworkers applied different physical treatments (i.e., drying, sonication 30/60 min, reflux 60/90 °C, maceration in MeOH) to determine their effects on *Spirulina* sp. amino acid profile, which they concluded is rich in amino acids related to umami flavor (i.e., Asp and Glu). Drying and methanol maceration showed to be the treatment that delivered the highest (8.37 g/100 g) and lowest (2.34 g/100 g) contents of Glu, respectively [231].

Campanella and coworkers assayed total and free amino acids from *Spirulina* sp.; they found that freshwater *Spirulina* contained relatively high concentrations of non-essential amino acids. The authors indicate that the samples tested were lysine-rich and limited in sulfur-containing amino acids. Free amino acids constitute as high as 2% of the amino acid input. Method-wise, the authors used an oxidation-capable acid, this is chancy as it may contribute to analytes deterioration. Additionally, the mobile phase already included the derivatization agent [219].

Al-Dhabi and Arasu quantified polyunsaturated fatty acids, sugars, polyphenol and total and free amino acids in *Spirulina* sp. In contrast to the authors mentioned above, this research group used pre-column derivatization and a dedicated column for analysis. Total amino acids contents ranged from 11.49 to 56.14 mg/100 g; from which essential amino acids accounted for 17.00 to 39.18%. [220].

3.3.3. Compound Feedstuff

For the specific case of feed, a time-reduced (i.e., complete separation in an eight-minute chromatographic run) analysis has been recently developed [225]. AOAC OMA^{SM} includes two different assays to determine amino acids based on LC; 992.12 design for pet foods using fluorescence and 999.13 include ninhydrin/Orto-phthalaldehyde (OPA) fluorescence or pulsed coulometric detection. Finally, a report made by Wang and coworkers described a successful simultaneous analysis of 20 amino acids without using derivatization using an evaporative light scattering detector [232]. More recently, underivatized amino acids have also been monitored using hydrophilic interaction liquid chromatography coupled with tandem mass spectrometry [233]. Herein, we included some examples of derivatization agents. However, we suggest the reader access a paper written by Masuda and Dohmae, which not only cites the four most commonly used reagents for amino acid derivatization, but also identifies their strengths and weaknesses [234].

3.3.4. Bacterial Cell Walls, Peptidoglycan, and Food-Extracted Peptides

A less common application for LC, is to monitor the products from the hydrolysis of bacterial cell walls (using enzymatic physical, and chemical approaches) and posterior fragment analysis. Desmarais and coworkers design a method that included the digestion of Braun's lipoprotein. Muropeptide fragments (monomers-trimers), 3,3-diaminopimelic acid among others [226]. Kühner and coworkers developed a similar application; complete muropeptide hydrolysis was accomplished within 24 h. UPLC/MS was used to monitor fragments. After BH_4^- reduction, both Gram-positive to Gram-negative bacteria can be evaluated after gradient modification [227]. In this regard, MSD (Mass Spectrometry Detectors) serve as a good reference for additional mass information, which will ease the peptidoglycan *in silico* reconstitution. This application has not found accommodation in food or feed, but it can correctly be adapted for bioreactions/fermentations or lactic bacteria.

Other applications for LC include, for example, the work by Marseglia and coworkers. They identified $n = 44$ different peptides from cocoa beans. The peptide fragmentation pattern in fermented cocoa samples was used to describe the geographic origin, different fermentation levels, and roasting. Vicilin, a storage protein, was identified in cocoa bean samples, information that can be useful to understand the biological activity of cocoa and to determine the aroma relevant peptides [228]. MS assisted analysis is advantageous as amino acids lack any distinctive chromophores and already have readily ionizable moieties. Prados and coworkers recently have described a method to isolate, characterize and identify peptides that can downregulate adipogenesis. The authors also used semipreparative fractionation to achieve the initial peptide separation [229].

3.3.5. Method Application Experience

When facing fresh feed products (e.g., wet pet food, forages) additional operation units such as lyophilization is necessary before sample treatment (see, for example, [235]). To obtain individual amino acids, most applications require acid or alkaline hydrolysis. However, amino acids are extremely susceptible to oxygen during hydrolysis, to prevent quantitative losses, we recommend the sample hydrolysis steps suggested elsewhere for furosine [236]. Additionally, pyrogallol in 1 g/100 mL is also used as a radical receptor (i.e., a radical sink) to avoid amino acid degradation. Particularly, Trp, Thr, and Tyr are usually lost during acid hydrolysis, cysteine is oxidized to cysteic acid, and asparagine and glutamine (if present) will transform to their respective acids. Hydrolysis may be performed using a feed of known concentration in parallel as a reference [237].

From the sample preparation standpoint, we have applied a Supelco ENVI-Carb SPE cartridge for cleanup as hydrolysate retain undesired particulates. A translucent extract is obtained after SPE that will be suitable for both FLD (Fluorescence Detector) and UV-Vis (Ultraviolet-Visible) detection-based analysis. Also, cleaner chromatograms are obtained as interferences are significantly reduced. This cartridge will adsorb (including those responsible for coloration) a great range of molecules, while the (charged) amino acids will not be retained. Sodium azide is applied routinely for extract storage. However, best results are obtained when measurements are performed immediately after preparation steps.

We have used a method based on OPA pre-column derivatization adapted from an established method for biopharmaceuticals [238]. We also recommend the strict use of a C_{18} guard column to increase column lifespan. When applied to feed samples and feed ingredients, essential amino acids covered include Arg, His, Ile, Leu, Lys, Met, Phe Val, and non-essential Ala, Asp/Asn, Glx, Cys/CY2, Gly, and Ser for a total of 14 amino acids. OPA derivatization is only effective under alkaline conditions (usually performed using borate buffer pH 8–10). Therefore, the feed hydrolysate must be neutralized (to pH 7.0) before injection, as the buffer will not be able to compensate for the [H_3O^+] that results from the acid treatment. Furthermore, 9-fluorenylmethyl chloroformate (FMOC) must be included during derivatization (additional to OPA) to obtain proline and hydroxyproline amino acids (see, for example, [224]).

Method automatization (when an automatic sampler is available) can concede an advantage since the reaction occurs in situ within the needle. Automated precolumn derivatization is also useful for unstable adducts (e.g., OPA derivatives). A benefit of amino acid derivatization is that most adducts can be monitored using a UV/VWD (Ultraviolet/Variable Wavelength Detector) or DAD/PDA (Diode-Array Detection/Photodiode-Array Detection), so even if the fluorescence detector is not available, analysis can still be performed. Though, the fluorescent detector can filter interferences, begetting cleaner chromatographs. We have also used the same method to assess the purity of feed grade amino acids, and taurine. The technique can also be applied to energy drinks to evaluate taurine in as a very simple "dilute and shoot" method after sonication for sample degassing. Interestingly, ninhydrin and OPA can detect complementary analytes to methods based in ninhydrin (see, for example, [223]).

3.4. Triphenylmethane Dyes

Malachite green is a dye usually used in aquaculture as a fungicide and antiparasitic due to its low cost and effectiveness [239]. The widespread use of this substance is not without downsides, though, including residue accumulation in fish tissue and contamination of sediments and water bodies, which can affect non-target organisms downstream (see, for example, [240,241]).

Recent and improved methods have found acceptance to monitor these kinds of dyes in fish tissue. For example, Hidaya and coworkers already conducted a short review on techniques available for detection of malachite and leucomalachite green in the fish industry [242].

Within, this paper, several LC-based techniques are mentioned. Triphenylmethane dyes suffer from reversible redox reactions; each form can be oxidized or reduced to one another (see, for example, [243]; Figure 5).

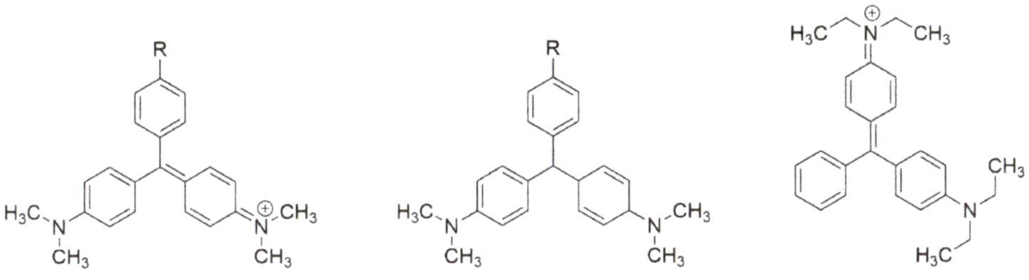

Figure 5. Chemical structures for three triphenylmethane dyes which are sharing a common phenyl backbone sharing a methylidyne. Each molecule has extended π-delocalized electrons justifying their crystal coloration and visible light absorption (ca. 621 nm for malachite green).

Table 11 shows a summary of various methods for the extraction and identification of malachite green and its metabolites. Although it is a common contaminant in aquaculture production, and research focuses on fresh residues from aquaculture production animals (fish, shrimp, lobster, among others), the development of methods should also be extended to the analysis of feed [244] as fish and shrimp feed are made from marine by-products. Doses on fish or shrimp range from 0.05–0.2 mg L^{-1} as an active ingredient have been used. Treatments for fish eggs include dosages of 5 mg L^{-1} is usually suggested for fish tanks. Laboratories usually measure malachite green with equipment able to detect tissue residues below 2 µg kg^{-1} (maximum permitted residue limit in fish tissue; [250]). A very interesting approach was made by Furusawa who developed a green chemistry method for malachite green and its metabolite [251].

Table 11. Measurement techniques meant for triphenylmethane dyes.

Matrix	Extraction Method	Measurement Method, Chromatographic Column	Reference
Fresh fish muscles	Extraction with 0.1 mol L^{-1} $NH_4O_2C_2H_3$ buffer, pH 4.5, HAH solution 0.25 g mL^{-1}, 1 mol L^{-1} p-TSA solution and can	LC: Cloversil-C_{18} 250 × 4.6 mm, 5 µm. Identification: MS/MS/ESI$^+$	[243]
Channel Catfish muscle	Extraction with McIlvaine buffer, TSA, and TMPD. Oasis MCX SPE columns	LC: Prodigy ODS-3 C_{18} 150 × 4.6 mm, 3 µm. Identification: MS/MS/ESI$^+$	[244]
Aquaculture water	Not indicated	LC: Phenomenex C_{18} 140, 250 × 4.6 mm, 5 µm. Identification: UV 558 nm (malachite green and crystal violet), FLD λ_{ex} 265, λ_{em} 360 nm (leuco forms)	[245]
Processed fish products	Extraction with ammonium acetate buffer, HAH solution, p-TsOH solution, and can	LC: 250 × 4.6 mm, 5 µm Capcell PAK C_{18} Identification: MS/MS/ESI$^+$	[246]
Fish tissue	Modified QuEChERS Extraction: NH_4O_2CH and can	LC: XCharge C_{18} column	[247]
Salmon	Extraction $C_2H_3O_2^-$ buffer, p-TSA solution, hydroxylamine and can	YMC phenyl 3-4-5 50 × 4.0 mm, 3 µm Identification: LC-MS/ESI$^+$/APCI	[248]
Fish feed	Extraction with ACN/CH_3OH/CH_3COOH	Chromolith® Performance RP-18e (100 × 4.6 mm) Identification: MS/MS/ESI$^+$	[249]

As previously mentioned, Wang and coworkers used solid-phase microextraction with the excellent result to assess malachite green, crystal violet and their respective metabolites using a monolithic fiber [245]. Bae Lee and coworkers homogenized fish tissue samples, and the extracted residues were partitioned into dichloromethane and an in situ oxidation with 2,3-dichloro-5,6-dicyano-1,4-benzoquinone. Afterward, cleaned-up was performed on neutral alumina and propyl sulfonic acid cation-exchange solid-phase extraction cartridges. Malachite green and crystal violet were determined at 618 and 588 nm using HPLC-Vis detector [246]. A common approach included analyzing dyes using traditional detectors and adding a step that included confirmation by

MS. Chengyun and coworkers relied on Oasis® MCX (a strong cation exchange-based adsorbent) to perform clean-up. After a two-step, QuEChERS extraction, dispersive solid phase extraction coupled with, both, a reverse phase and strong anion exchange (as well as a mixed mode adsorbent) cleanup was tested. Residues of the dyes were evaluated in codfish [247]. However, we do not see how anion exchange favors dye-stationary phase interaction, since all parent compounds are positively charged. Noteworthy, usually reverse phase columns can resolve these types of dyes with ease, even if several analytes are to be evaluated simultaneously. Croatia and Iran are specific examples of countries which have stated have found residues of this dye in fish tissue [252,253]. Both cases demonstrate the need to assess these compounds in food items. However, both research groups used immunoassays to evaluate the contaminant. AOAC method OMA^{SM} 2012.25 is a reference based on LC-MS/MS to assess triphenylmethane dyes and their metabolites in aquaculture products.

Additionally, US FDA reference method is based on the isolation of malachite green using alumina/propyl sulfonic solid phase extraction cartridges previous to Non-Discharge Atmospheric Pressure Chemical Ionization and an LC-MS^n; quantification was performed in salmon [248]. Finally, since fish and shrimp compound feed can also be based in aquaculture by-product meal, as a source of protein, contaminated tissue can reach the final product. Hence, the need for feed analysis is evident, as it shows, Abro and coworkers [249].

4. The Common Ground among Measurements Performed in Food and Feed Laboratories

4.1. Nitrates and Nitrites

Nitrates and nitrites are natural compounds that are part of the nitrogen cycle, but especially high dosages of these ions are registered because of anthropological activities [254,255]; they enter human diets by means of drinking water, leafy vegetables, and cured meats. Noteworthy, these ions have been authorized as additives in several countries including the European Community [256,257].

Though there is evidence that both ions have a relevant biological and physiological function, special attention has been paid to nitrates and nitrites and their metabolites such as *N*-nitrosamines and nitrous oxide as all these molecules may pose a health hazard [256,257]. For example, these compounds have been related to colorectal cancer [256–261]. Hence, risk management and assessment in food have been proved necessary [258]. Regarding the quantification of NO_2^- and NO_3^- using HPLC, there are two main approaches used i.e., ion exchange and reverse phase columns (Table 12).

4.1.1. Ion Exchange Chromatography

When analyzing crops, one must consider that cultivar, and harvest date can affect the nitrate levels of selected vegetables. Hence, maximum levels have been set by European legislation accordingly [262]. For example, Brkić and coworkers analyzed several leafy greens (*n* = 200) in two different seasons, in order to evaluate differences in ion content and encountered considerable differences among vegetable and sampling season [263]. Pardo-Marin and coworkers assessed vegetable-based baby foods, considering the levels found within these types of foods. They calculated ion ingestion between 13–18% of the acceptable daily intake for an infant. [264]. Quijano and coworkers assessed the nitrate content of vegetables (*n* = 533); they obtained values up to 3509 mg kg^{-1} in chard samples. They calculated an intake of 490 mg kg^{-1} bw day^{-1} for a young population, values which tend to increase the risk of exceeding acceptable intake values [265]. The main advantages in using ion exchange columns are that the separation can be accomplished using aqueous buffers which are made up from relatively cheap salts, making the methods apt for green chemistry and avoid mobile phase drift [263].

4.1.2. Ion Pairing and Reverse Phase Chromatography

Tetrabutylammonium salt has also been used as an ion-pairing agent coupled with reverse phase columns (Table 12). For example, Hsu and coworkers used a reverse phase approach to assess both ions in cured meats and vegetables. The authors found the highest values of NO_3^- in spinach (4849.6 mg kg^{-1}) and for NO_2^- in hot dogs (78.6 mg kg^{-1}) [266]. Nitrite tends to oxidate to NO_3^-, the authors cite several factors affecting nitrate and nitrite recovery in foods (e.g., temperature, pH, metals). Usually, non-desired compounds found in greens differ from those found in meat products, for which proteins interfere significantly. Meat sample extracts will need pH adjustments and higher temperatures are needed to improve recovery. Some of these parameters must be monitored during analysis, especially when vegetables are subject of study [266]. Croituru used a similar approach to assess human, rabbit, rat urine as well as vegetables. However, they roduced adducts (an azo dye, $HO_3SC_6H_4-N=N-C_{10}H_6NH_2$) based on Greiss reaction (sulfanilic acid form a diazonium cation ($HO_3SC_6H_4-N\equiv N^+$) with NO_2^- and then with 1-naphthylamine) for NO_2^- that was measured at 520 nm [267,268]. Interestingly, the authors followed the reaction with mass spectrometry. We encourage the reader to pay special attention to this paper as highlights difficulties during method development. The author concluded that while useful, the use of Greiss reaction, spectrophotometrically, is unadvisable as several samples may exhibit additional confounding compounds that may behave similarly as the NO_2^- ion adduct. However, is quite valuable as a derivatizing agent when coupled with HPLC; the method can work with samples of different origins without the need for further modifications [267]. Samples were decolorized with carbon and $ZnSO_4$ was applied for protein precipitation to overcoming this matrix interference and enhance the sensitivity. Croituru and coworkers used a validated method to assess NO_2^- and NO_3^- in vegetables for self-consumption; toxicologically speaking, the NO_2^- content found in the samples was deemed too low to represent a hazard [269].

Stationary phases containing only alkyl chains have been used, but it is also possible to find mixed stationary phases, for example, Abdulkair and coworkers assayed NO_2^- and NO_3^- using a stationary phase containing both alkyl groups and phenyl groups (Table 12) to separate both ions successfully after sonication [270].

Chou and coworkers assessed both ions in vegetables and observed a high concentration variability was observed which reflect differences in environmental conditions [271]. The authors also optimized critical chromatographic parameters such as pH, organic solvent fraction, and flow [271]. In this regard, the methanol fraction optimization was demonstrated to be paramount to improve octylammonium solubility and achieve an optimal resolution between both ions. In contrast, pH and flow variations tend to have an effect only on chromatographic run times and not so much in resolution.

Table 12. Common chromatographic approaches for the determination of nitrate and nitrite ion.

Matrix	Mobile Phase Composition		Measurement Method, Chromatographic Column	Reference
Ion Exchange Chromatography				
Leafy greens	10 g L^{-1} of KH$_2$PO$_4$, pH 3.0		Waters IC-PAK HC anion exchanger (150 × 4.6 mm), UV λ 214 nm	[263]
Baby foods	Phosphate 5 mmol L^{-1} (pH 6.5)		Waters IC-PAK HC anion exchanger (150 × 4.6 mm), UV λ 214 nm	[264]
Vegetables	Phosphate 5 mmol L^{-1} (pH 6.5)		Waters IC-PAK HC anion exchanger 150 × 4.6 mm, 10 μm, UV λ 214 nm	[265]
Reverse Phase Chromatography				
Matrix	Ion pair reagent	Mobile phase composition	Measurement method, chromatographic column	Reference
Cured meat and vegetables	Tetrabutyl ammonium (TBA)	MeOH:H$_2$O (75:25)	Phenomenex C$_{18}$ 110 Å Gemini 250 × 4.6 mm, 5 μm. PDA λ 214 nm	[266]
Vegetables	TBA, Greiss reagent	Gradient MeOH/ACN/H$_2$O	X Bridge C$_{18}$, 50 × 2.1 mm, 2.5 μm. UV-Vis λ 222 (nitrate) and 520 nm (nitrite)	[267]
Cured meats	3 mmol L^{-1} TBA	ACN/2 mmol L^{-1} HPO$_4^{2-}$ pH 4	RP-thermophenyl hexyl, 150 × 4.6 mm, 3 μm, UV λ 205 nm	[270]
Vegetables	0.1 mol L^{-1} octyl ammonium salt	OA buffer pH 7.0/MeOH (70:30)	Phenomenex Luna C$_{18}$ 250 × 4.6 mm, 5 μm, UV λ 213 nm	[271]
Dried vegetables and water	Triethylamine (TEA)	C$_6$H$_{13}$SO$_3$H, H$_2$PO$_4^-$, TEA pH 3.0/MeOH (80:20)	C$_{13}$ 250 × 4.6 mm, 5 μm, UV λ 222 nm.	[272]
Ham	n-octylamine/TBA	0.01 mol L^{-1} n-octylamine/5 mmol L^{-1} TBA pH 6.5	Acclaim™ Polar Advantage and C$_{18}$ Thermo Scientific™, HyPURITY™, 250 × 4.6 mm, 5 μm	[273]

4.1.3. Miscellaneous Methods for Nitrates and Nitrites

In contrast with ion pairing approaches, dos Santos and coworkers developed a method based on the reaction of the NO$_2^-$ with 2,3-diaminonaphthalene to yield a highly specific fluorescent 2,3-naphthotriazole adduct (λ$_{ex}$ 375 λ$_{em}$ 415 nm), under acidic conditions, to assess the ions in beetroot [274]. Cassanova and coworkers have developed an application for HPLC derivatization based on VCl$_3$, 4-nitroaniline, methanesulfonic acid, and N-(1-naphthyl)-ethylenediamine. Under these conditions, a post-column reduction of nitrate to nitrite can be accomplished [275].

4.1.4. Method Application Experience

The preferred methodology used in our laboratories is based on the chromatographic determination of NO$_2^-$ and NO$_3^-$ anions simultaneously. Reverse phase (using a C$_{18}$ column, i.e., Zorbax Eclipse 5.0 μm, 4.6 mm × 150 mm, set at 30 °C and 0.6 mL min^{-1}) HPLC-PDA or -VWD (213 nm as the absorption spectra maximum) is sufficient to perform the assay [266,271]. It is important to emphasize that for the detection and separation of inorganic anions, in this case NO$_3^-$ and NO$_2^-$, the mobile phase must contain a complementary counter ion that interacts with it and with the bonded stationary phase of the column concurrently. In the absence of the counter ion, no interaction with the column is achieved and, as a result, no retention will be obtained at all, as the ions would come out in the void. In this scenario, a tetrabutylammonium salt (e.g., tetrabutylammonium hydrogen sulfate, TBAHS, 155837 Sigma-Aldrich) is a possibility (Figure 6B). In this case, the four alkyl chains from the reagent interact with the eighteen-carbon alkyl chains of the stationary phase and, at the same time, with the NO$_2^-$/NO$_3^-$. The elution order may be explained by considering a more delocalized negative charge (among three oxygen atoms) in NO$_3^-$ and the bent geometry of NO$_2^-$ due to the nitrogen atom-containing an electron lone pair. Interestingly, NO$_2^-$ is a larger anion (0.192 nm), when hydrated, than NO$_3^-$ (0.179 nm) [276]. Now, depending on the length of the column, the affinity of the this will not be sufficient to resolve peaks from the solvent front (specially the first peak; NO$_2^-$), this issue is easily solved including acetonitrile in the mobile phase, using slower flows, a longer column or even an ion pair agent with longer alkyl chains (e.g., octylamine). The mobile phase used is 20% acetonitrile,

80% TBAHS 5 mmol L^{-1}, at a 6.5 pH. Interestingly, when injecting a solution with both ions present and at the same concentration, the response (the signals obtained), is very similar in area/height and, as such, sensitivity is very close for both anions.

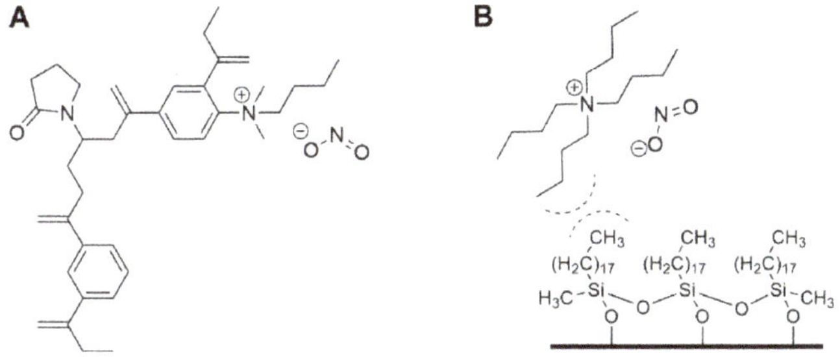

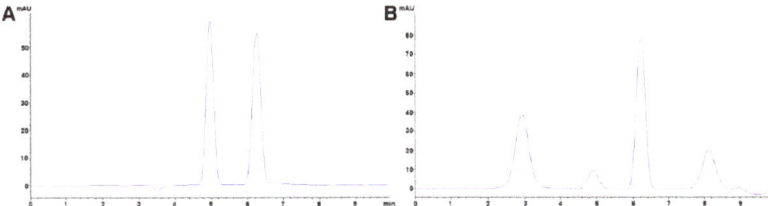

Figure 6. Schematic representation for the interaction of nitrite ion with (**A**) a cation exchange stationary phase or (**B**) interaction with TBAHS present in the mobile phase and stationary phase C$_{18}$.

The same methodology has been used in feed to assay hay samples (Figure 7A,B) that were presumed as the source of intoxication in horses [277]. In this case, from ten samples assayed, three (average concentrations of 92.77 ± 60.88 mg kg^{-1}) and six (average concentrations of 92.13 ± 47.55 mg kg^{-1}) samples tested positive for NO$_2^-$ and NO$_3^-$, respectively (unpublished data). Forage and swine compound feed samples (n = 10) have also been assayed with this method obtaining values from <5 to 23.69 and 2.30 to 4.96 and 925.15 to 1135.10 and 989.51 to 1479.71 mg kg^{-1} for NO$_2^-$ and NO$_3^-$, respectively on both accounts. In the case of feeds and fish meals, which suffer from severe matrix effects, SPE has been applied, with good results, as a cleanup and concentration step. Specifically, Oasis® MAX cartridges, conditioned with 2 mL methanol, and 4 mL water, load 1 mL sample, wash 3 × 1 mL water, elute with 2 mL 0.5 mol L^{-1} NaCl solution. Chromatograms improve drastically when the elution from the cartridge is performed using the mobile phase.

Figure 7. Chromatograph of (**A**) an aqueous 10 mg L^{-1} nitrite (4.95 min) and nitrate (6.26 min) standard (**B**) hay sample after extraction with hot water, SPE cleanup, and micropore filtration presence of nitrite (4.91 min) and nitrate (6.23 min) is evident.

4.1.5. Legislation

Regulation 2002/EC/32 sets limits for NO$_2^-$ in fish meal (i.e., 60 mg NaNO$_2$ kg^{-1}) and complete feedingstuffs (i.e., 15 mg NaNO$_2$ kg^{-1}) excluding those intended for pets except birds and aquarium fish. We refer the reader to two thorough reviews that tackle regulatory as well as methodological topics [278,279].

4.2. Carotenoids

Chemically, carotenoids are conjugated hydrocarbons that may be further classified as carotenes (without any oxygen molecules) and xanthophylls (with one or more oxygen molecules). Carotenoids are widespread natural pigments, are recognizable from the bright colors (yellow, orange, red, or purple) that they often confer on plant and animal organ. The molecules responsible for producing said coloration must be attained from dietary sources. For example, lutein and zeaxanthin are carotenoid pigments that impart yellow or orange color to various common foods such as cantaloupe, pasta, corn, carrots, orange/yellow peppers, fish, salmon and eggs, β-carotenoid and isomer are found in sweet potatoes, dark leafy greens, butternut squash, lettuce, red bell peppers, apricots, broccoli, and peas, and lycopene are in tomato. As molecules with a conjugated double bond system, carotenoids serve several physiological functions (e.g., antioxidants, immunostimulants, photoprotection, visual tuning, among others). This electron delocation causes them to be particularly unstable compounds, especially sensitive to light, heat, oxygen, and acids. Hence, several precautions have been taken while extracting carotenoids. For example, must be carried out in dim lighting; use rotary evaporation at low temperature and reduced pressure also it has to be carried out under a stream of nitrogen. Finally, samples should be stored in the dark, at about −20 °C [280,281].

Carotenoids are fat soluble but, because of the high moisture content of plant tissues, a preliminary extraction solvent miscible with water (e.g., methanol or ethanol) is generally necessary to allow for penetration of the extraction solvent.

Saponification is required to remove interference as neutral fats, chlorophylls, and chlorophyll derivatives. Usually, this procedure is carried out with potassium hydroxide in methanol. Then, it is necessary to perform liquid–liquid extraction using a water-immiscible solvent (e.g., ethyl acetate, ethyl ether, hexane) to obtain the unsaponifiable fraction, where carotenoids should be present. [280–298]. The identification and quantification require high-resolution techniques; the reversed-phase high-performance liquid chromatography has been used routinely to determine carotenoids because of its satisfactory separation efficiency. So, several factors must be evaluated to employ HPLC technique such as column type, mobile phase, chromatographic conditions. Several methods for carotenoid analysis are summarized in Table 13.

Regarding column type, the analysis can be performed using a C_{18} column. However, YMC C_{30} Carotenoid dedicated column provides excellent results, had better resolution than a C_{18} column for separation of carotenoids and their geometric isomers. The thirty-carbon alkyl chains interaction with the carotenoid lipophilic profile guarantee less peak distortion and better resolution [280,281]. Compounds such as α/β/γ/δ/ε-carotene, lutein, zeaxanthin, β-cryptoxanthin, dehydrolutein, anhydrolutein, astaxanthin, galloxanthin, α-doradexanthin, adonirubin, and canthaxanthin can all be separated using the aforementioned chromatographic column.

According to Huck and coworkers, the flow rate did not significantly influence the resolution, but it is essential to use an adequate flow to generate acceptable column back pressure. Also, they studied the effect of column temperature on the separation of lutein, zeaxanthin, β-cryptoxanthin, and β-carotene. The column temperature was varied between 21 and 80 °C; the best selectivity being achieved at 21 °C, at a temperature of 34 °C, zeaxanthin could not be easily separated from lutein. The authors concluded that maintaining a constant temperature during carotenoid analysis is critical as small changes in the ambient temperature can cause significant changes in the chromatographic selectivity of the carotenoids and at temperatures higher than 60 °C, the investigated carotenoids unstable.

In the case of the mobile phase, the same authors indicated that carotenoid selectivity was better using tetrahydrofuran, rather than ethyl acetate, and also better than MeOH and ACN. Carotenoids are sensitive to degradation on the stationary phase of the column by the presence of silanol groups.

Table 13. Common chromatographic approaches for the determination of carotenoids.

Matrix	Extraction Method	Measurement Method, Chromatographic Column	Reference
Camu–camu (*Myrciaria dubia* (Kunth) Macvaugh)	Extracted from the crushed peel with acetone transferred to petroleum ether/diethyl ether and saponified with 10% KOH methanolic	HPLC-PDA Quantitative: C_{18} Nova-Pak ODS 300 × 3.9 mm, 4 µm set at 29 °C, mobile phase ACN/H_2O/ethyl acetate For Qualitative: C_{30} YMC Carotenoid 250 × 4.6 mm, 3 µm at 33 °C. Mobile phase MeOH/MTBE (methyl tert-butyl ether)	[282]
Algae species, *Chlorella vulgaris*, and *Scenedesmus regularis*	Extraction with n-hexane–EtOH–acetone–toluene (10:6:7:7) 1 h, Saponification: 40 g/100 mL methanolic KOH at 25 °C in the dark for 16 h	PDA, YMC Carotenoid (250 × 4.6 mm, 5 µm, MeOH/ACN/H_2O (84:14:2) and CH_2Cl_2 gradient UV λ 450 nm	[284]
Tissues of a species of colored bird (*Taeniopygia guttata*)	Plasma and liver extract: n-hexane:MTBE (1:1) Adipose tissue, retina, beak, legs: Saponification 0.02 mol L^{-1} methanolic KOH for 6 h, organic solvent extraction	PDA, YMC Carotenoid 250 × 4.6 mm, 5 µm, MeOH:ACN:CH_2Cl_2 linear gradient)	[285]
Taiwanese sweet potatoes (*Ipomoea batatas* (L.) Lam.)	Extraction with hexane/acetone/EtOH (2/1/1) containing $MgCO_3$ and BHT (butylated hydroxytoluene) by 0.5 h, Saponification with 40 g/100 mL methanolic KOH for 3 h under nitrogen gas at 25 °C	PDA, YMC Carotenoid 250 × 4.6 mm, 5 µm, at 25 °C, MeOH–ACN–H_2O (84:14:2) and CH_2Cl_2, UV λ 450 nm	[287]
Mashed orange-fleshed sweet potato	Extraction with acetone, THF, and THF:MeOH (1:1)	PDA Phenomenex LUNA C_{18} ODS, 250 × 4.6 mm, 5 µm, ACN:THF:MeOH: 1 g/100 mL $NH_4C_2H_3O_2$ (68:22:7:3) at room temperature and 450 nm	[288]
Selected vegetables	Extraction with THF and MeOH (1:1), petroleum ether containing 0.1 g/100 mL BHT and 50 mL 10 g/100 mL NaCl	HPLC-APCI$^±$–MS, Phenomenex Luna Si C_{18} column (250 × 2 mm, 5 µm), ACN 0.1 g/100 mL/MeOH (0.05 mL L^{-1} $NH_4C_2H_3O_2$, 0.05 mL/100 mL TEA)/$CHCl_3$ (0.1 g/100 mL BHT)/n-heptane (0.1 g/100 mL BHT), ambient temperature	[289]
Fresh and Processed Fruits and Vegetables	Extraction under subdued yellow light. 50:50 acetone/hexane Saponification: saturated methanolic KOH SPE Alumina N Sep Pak	UV, YMC Carotenoid 250 × 4.6 mm, 5 µm, mobile phase 89:11 MeOH/MTBE	[290]
Papaya (*Carica papaya* L., cv. Maradol)	Freeze-dried papaya homogenized in hexane: CH_2Cl_2 (1:1). Organic phase was separated and saponified with methanolic KOH 40 g/100 mL (1:1) for 1 h at 50 °C	HPLC-DAD-MS/MS-ESI$^+$, C_{30} YMC Carotenoid 250 × 4.6 mm, 3 µm, at 15 °C. The mobile phase MeOH and MTBE	[291]
Papaya (*Carica papaya* L.)	MeOH ethyl acetate, and light petroleum (bp 40–60 °C) containing 0.1 g/L of both BHT and BHA (butylated hydroxyanisole)	DAD, YMC Carotenoid 250 × 3.0 mm, 3 µm at 25 °C. Mobile phase MeOH/MTBE/H_2O (91:5:4) and MeOH/MTBE/H_2O (6:90:4)	[292]
Yellow and red nance fruits (*Byrsonima crassifolia* (L.) Kunth)	Sample with $CaCO_3$, NaCl solution (30 g/100 mL, were extracted MeOH/ethyl acetate/light petroleum 1:1:1 Saponification: methanolic KOH 30 g/100 mL stirring for 23 h	HPLC-PDA-MS/ESI$^±$, C_{30} YMC Carotenoid 250 × 3.0 mm, 3 µm, MeOH/H_2O/aqueous $NH_4C_2H_3O_2$ 1 mol L^{-1} (90:8:2) and MTBE/methanol/aqueous $NH_4C_2H_3O_2$ 1 mol L^{-1} (78:20:2), gradient, at 40 °C. UV λ 450 nm	[295]
Red and Yellow Physalis (*Physalis alkekengi* L. and *P. pubescens* L.) Fruits and Calyces	Extraction: $CaCO_3$, MeOH/ethyl acetate/petroleum ether (1:1:1), containing 0.1 g L^{-1} BHA and BHT., sonication	HPLC-PDA-MS/ESI$^±$, C_{30} YMC Carotenoid (250 × 4.6 mm, 3 µm, mobile phase MeOH/MTBE/H_2O (80:18:2) and MeOH/MTBE/H_2O (8:90:2), both 0.4 g L^{-1} $NH_4C_2H_3O_2$.Gradient.	[297]

Solvent modifiers could be added to the mobile phase, for example triethylamine (TEA). Free silanol groups on the surface of silica deprotonate in the presence of the basic molecules, preventing the analyte from interacting with the medium. The TEA generates a positive impact on peak symmetry, reducing the peak tailing effect, reducing the retention time, and improving the recovery. The addition of triethylamine to the mobile phase can also have negative consequences, such as changes in the pH of the mobile phase; therefore, it is recommended that TEA be used in low concentration (less than 0.05 mL/100 mL) [299].

When using chlorinated solvents, the addition of ammonium acetate to the MeOH provides sufficient buffer capacity to prevent losses due to acid degradation of carotenoids. Some papers use MTBE as part of the mobile phase. The advantage in using this solvent, instead of chlorinated solvents, lies in the MTBE is less volatile (55.2 vs. 39.6 °C, respectively) and less toxigenic. Depending on the solvent system, a good compound separation may require a longer run time and poorer resolution compared with MeOH/ACN/H_2O/CH_2Cl_2. Carotenoid content in tropical pigment-bearing fruits [281,295,300–302], and fish [302] have also been described.

Method Application Experience

The preferred methodology used in our laboratories is based on the work by Gayosso and coworkers with some modifications [282]. We use MTBE/MeOH as the mobile phase with a gradient system for 45 min with YMC C$_{30}$ (150 × 3.0 mm, 3 μm) at 0.6 mL min^{-1} and 30 °C. These conditions were applied to identify and quantify carotenoids in food matrices such as palm oil, peach palm, sweet potatoes, papaya, and guava. We extracted the carotenoids from these matrices using a saponification procedure, followed by extraction with ethyl ether. This solvent evaporates at 40 °C and the residue is reconstituted in CHCl$_3$. Undesired coextractants (e.g., waxes, sterol and tocophenol esters) are usually better solubilized with this solvent than MTBE saving from additional filtration steps and within-system precipitation. Optimization of injection volumes and initial composition of the mobile phase can somewhat mitigate the effect that injecting in a different solvent [303]. Analogous to polyphenols, carotenoid extraction methods must contemplate ester hydrolysis or other treatments to ensure the quantitation of overall amounts of carotenoids. For example, it is common to find carotenoid esters in food matrices, and these adducts present several intrinsic difficulties during carotenoid determination [295]. However, mass spectrometry-based LC is a powerful tool able to discriminate both parent compounds and their esters [295]. Recently, Wen and coworkers identified n = 69 carotenoids esters in *Physalis alkekengi* L. and *P. pubescens* L. fruits [297]. Additionally, BHA and BHT are common organic-solvent-soluble antioxidants to preserve carotenoid integrity [298]. Finally, our laboratory has also assessed carotenoid content in plasma from colored tropical frogs (*Agalychnis callidryas*).

4.3. Carbohydrates and Sugars Soluble in Ethanol

Animal feeds are, by definition, based on vegetable/plant sources that use carbohydrates as storage compounds, structure elements, and energy sources [10]. Then, carbohydrates form the most substantial portion of the organic matter in feeds; they can be divided into two main categories non-structural and structural carbohydrates. We encourage the reader to examine an excellent review of carbohydrate and organic acid in food commodities intended for human consumption by da Costa and Conte-Junior [304]. A great starting point for reviewing different approaches for carbohydrate analysis is the thesis written by de Goeij [305].

4.3.1. Carbohydrate Measurement Using Amino-Based Columns

Xu and coworkers compared two methods for sample cleanup and extraction. A macroporous resin was compared to a solid phase sorbent based on alkyl chain. From the two approaches, SPE showed less analyte loss (11.32 vs. 0.69%). However, the discoloration ratio was similar for both methods. Sugar profile from molasses samples was obtained [305] after pigments, nitrogen compounds, and inorganic ions were removed. The analysis was performed using two NH$_2$-based columns. Under the same conditions, it was concluded that the Zorbax Carbohydrate column showed better performance. Agius and coworkers recently developed a method to determine organic acid and sugars in tomato fruits [306]; the authors used ACN to improve peak shape. RID (Refractive Index Detector) is used for carbohydrate analysis since sugars do not have chromophores and alternative detectors (e.g., MS) are expensive. RID is the detector of choice in many labs for sugar profiling (Table 14) despite its relative lack of sensitivity. However, usual concentrations found in fruits counteract the issue.

Table 14. Different methods and stationary phases to assess carbohydrates in food matrixes.

Matrix	Sample Pretreatment, Extraction	Mobile Phase Composition	Measurement Method, Chromatographic Column	Reference
Amine-Based Columns				
Molasses	1. SPE Sep-Pak C_{18} 2. Microporous resin discoloration (Seplite LX), 30 °C	ACN:H_2O (75:25)	RID, IS: maltose 1. Zorbax Carbohydrate 2. Ultimate^{TM} XB-NH_2; both 250 × 4.6 mm, 5 µm	[307]
Tomato	Filtration, ACN/H_2O (45:55), SPE Chromabond NH_2	ACN:H_2O (80:20)	Nucleodur 100-5 NH_2 125 × 4 mm, RID, IS: lactose	[306]
Amide-Based Columns				
Confectionery, chocolate products, snacks	Defat (when applicable), H_2O 80 °C (+EtOH for chocolate products)	gradient ACN/H_2O + 0.05 mL/100 mL ethanol- and triethyl- amine	UPLC-ELSD, Acquity BEH Amide (50, 100, 150) × 2.1 mm, 1.7 µm, 85 °C	[308]
Apple Juice	Filtration	H_2O	RID, Sugar Pak^{TM} 300 × 6.5 mm, 10 µm, 80 °C	[309]
Ligand-Based Columns				
Tubers	1. H_2O 92 °C 2. Reflux MeOH/H_2O (50:50) 3. Activated Charcoal/MeOH 4. 2×·MeOH/H_2O 50:50 SPE Bond-Elut C_{18}	10 mmol L^{-1} H_2SO_4	RID, UHPLC Aminex HPX 87H 300 × 7.8 mm, 9 µm, 18 °C	[310]
Foods	Liquids: H_2O/EtOH (50:50) Solids: H_2O 65 °C, sonication, +EtOH Fat-/Protein-rich: H_2O/EtOH (20:80), sonication	1. H_2O 2. ACN:H_2O (9:1)	1. HPLC-RID Ultron PS-80P 300 × 6.5 mm, 10 µm, 50 °C 2. LC-MS-ESI$^±$ Unison UK-Amino 150 × 2.0 mm, 3 µm	[311]
Derivatization-Based Approaches				
Fruit tree buds	MeOH extraction, benzyl alcohol/NaOH 8 mol L^{-1}, SPE C_{18}	Gradient ACN/H_2O	PDA λ 228 and 248 nm, Shim-pack C_{18} 250 × 4.6 mm, 5 µm	[312]
Foods	2,3-naphtalenediamine, iodine, HAOc	$NH_4O_2CH_3$ 50 mmol L^{-1} pH 5.0 in ACN/MeOH (70:30)	RID (sucrose/fructose)/UV λ 310, FLD $λ_{ex}$ 320 $λ_{em}$ 360 nm, C_{18} 250 × 4.6 mm	[313]
Normal Phase				
Glycine max (L.) Merr	H_2O 55 °C, ACN	H_2O/ACN + Acetone (75:25)	ELSD, Prevail^{TM} Carbohydrate ES 250 × 4.6 mm, 5 µm	[314]
Complex Carbohydrates (e.g., Inulin and Fructans)				
Plants and feed materials	H_2O	H_2O or H_2SO_4 0.01 mol L^{-1}	1. Knauer Eurokat Pb 2. Nucleosil CHO 620 3. Nucleosil CHO 682 (Pb) 4. Biorad Aminex HPX-87C. All columns 300 × 7.8 mm, RID	[315]
Wheat	HCl 60 mmol L^{-1} 70 °C, quenching Na_2CO_3	90 mmol L^{-1} NaOH	Carbopac-PA-100 250 × 4.0 mm, PAD	[316]
Fungus sucrose fermentation	Filtration	ACN/0.04 g/100 mL NH_4OH (70:30)	RID, Knauer Eurospher 100-5 NH_2 Vertex 25 × 4.6 mm	[317]
Starch from feeds	heat stable amylase and amyloglucosidase	ACN/H_2O (80:20)	RID, Zorbax Carbohydrate 150 × 4.6 mm, 5 µm	[318]
Wine	Diluted 1:9 with EtOH 70 mL/100 mL and 1-phenyl-3-methyl-5-pyrazolone derivatization	ACN/0.1 mol L^{-1} g/100 mL $NH_4C_2H_3O_2$ (70:30)	UV λ 245 nm, Eclipse XDB-C_{18} 250 × 4.6 mm, 5 µm	[319]
Bacterial Exopolysaccharide	Microplate polysaccharide hydrolysis 4 mol L^{-1} CF_3COOH, 90 min at 121 °C. Derivatization 1-phenyl-3-methyl-5-pyrazolone	5 mmol L^{-1} $NH_4C_2H_3O_2$ pH 5.6/ACN gradient	Gravity C_{18}, 100 × 2 mm, 1.8 µm, HPLC-UV-Ion trap/ESI$^+$-MS	[320]

4.3.2. Carbohydrate Measurement Using Amide-Based Columns

Koh and coworkers developed a method using an amide-based column, which is designed to retain polar molecules [308]. Contrary to their amino counterparts, these columns can retain analytes wide range of mobile phase pH. Thirteen sugars were separated including monosaccharides, disaccharides, sugar alcohols. This separation is impressive since it includes several molecules commonly used as sugar substitutes or replacement sweeteners. Organic amines within the mobile phase are used as stationary phase modifiers [308]. The authors recommended the use of a 150 mm column as the reduction of time of analysis using shorter lengths, compromise resolution. However, peaks obtained on longer columns are typically wider peaks resulting in lower sensitivity due to increased diffusion.

4.3.3. Carbohydrate Measurement Using Ligand Exchange-Based Columns

Duarte-Delgado and coworkers assayed four different extraction methods for sucrose, glucose, and fructose, and demonstrated that a double aqueous MeOH extraction was the more efficient approach for the determination of these sugars [310]. The authors used SPE and guaranteed the removal less polar compounds and avoid possible co-elution with sugars during HPLC analysis. Extraction method seems to be more critical for mono than disaccharides, and starch gelification appears to be an interference when extraction is performed with hot water. Zielinski and coworkers a cation exchange gel in calcium form column to determine sucrose, D-glucose, fructose, and sorbitol in different ripe stages and during senescence of *Malus domestica* (Suckow) Borkh [309].

Senescent apple juice showed higher sugar concentration; a stage in which fruit is better suited for fermentation. Shindo and coworkers used recovered sugars from samples such as orange juice, yogurt, chewing gum, milk, and biscuits (this last matrix needed a triple extraction to obtain adequate recoveries). Additionally, the authors optimized column temperature and flow rate [311].

4.3.4. Reverse Phase Columns and Sugar Derivatization Techniques

Several detection systems are used to detect carbohydrate after chromatographic separation, an approach commonly used is the pulsed electrochemical detection. A thorough review of this technique has been already written by Corradini and coworkers [321]. Evaporative light scattering detector has also been used to assay sugars. Dvořáčkova and coworkers wrote a comprehensive review of this technique [322]. The most common detector for chromatographic analysis of sugars is refractive index. UV detection is usually inconvenient as the wavelength 210 nm (low range of the UV) has the disadvantage of exhibiting interferences. An easy way to circumvent this to derivatize using pyrazolones (e.g., 1-phenyl-3-methyl-5-pyrazolone) to form Schiff bases with reducing sugars and monitor using 248 nm. This approach only works for reducing sugars. Hence, sucrose will not be detectable. Additionally, a C_{18} column (usually readily available) can be used to separate the adducts. Canesin and coworkers analyzed sorbitol from lateral buds of fruit trees (e.g., black mulberry, peach, avocado, and pear) as a way to monitor primary photosynthesis products [312]. In this case, traditional detection systems are not useful as levels of sorbitol are in the µg per mg.

Hung and coworkers were able to add a fluorophore to aldol sugars assisting in their detection and mass fragmentation [313]. Naphthylimidazole fluorescent derivatives were obtained successfully for sugars (only for reducing aldoses) extracted from beverages such as fruit juice, yogurt, coffee drink, milk tea, and flavored milk. Additionally, oligosaccharides from a *Solanaceae* were identified using the approach above and NMR as an additional confirmatory tool. Recently, special attention has been drawn toward added sugars in food commodities; sterner regulations have been set in different countries due to population health concerns such as obesity, diabetes, and heart disease [323]. Hung and coworkers also used their approach to assess added sugar in the food items tested [313]. Carbohydrates analysis in food should contemplate, systematically, added sugars during chemical determinations [324].

4.3.5. Aqueous Normal Phase Chromatography for Sugars

Interestingly, Valliyodan and coworkers used an aqueous normal phase approach based on a hydrophilic polymeric gel) to assess sugars from soybean. The addition of just 20–30 mL/100 mL

4.3.6. Complex Carbohydrates and Conjugates

Hydrolysis of complex (mainly structural) carbohydrates has been used previously to assess them [325,326] by indirect determination of their basic units and building blocks. Several approaches can be used to achieve this [325]. However, HPLC can be an attractive one since it provides high specificity and selectivity. As hydrolysis usually produces considerable concentrations of the monomer, usually sensitivity is not an issue. For example, we have used endo-1,4-β-mannanase (EC 3.2.1.78) to break down and indirectly determine mannan, monitoring mannose. Similarly, hydrolysis can be used to assess the quality of commercial mannanase. Mannanase is commonly used as a feed ingredient to improve nutrient absorption [10]. Here, an enzyme of known activity (a standard, see for example E-BMANN from Megazyme) is directly compared to the commercial one (the feed additive); a galactomannan polysaccharide (like guar gum) can be used as the substrate.

Weiß and Alt described an exhaustive method to assay sugars in plant materials and feeds. Separation of the following was achieved: inulin, verbascose, stachyose, raffinose, cellobiose, sucrose, isomaltose, maltose, lactose, glucose, xylose, galactose, rhamnose, arabinose, fructose, mannose, ribose, and mannitol [315]. Flow rate, temperature, mobile phase composition, and injection volume were optimized. From the series of columns tested, the Nucleosil® Sugar 682 Pb (Macherey-Nagel GmbH & Co. KG, Düren, Germany) was finally used at 85 °C, H_2O at 0.4 mL min^{-1}, and using 20 µL. Recent data show that inulin-rich diets can benefit gut microbiome, notwithstanding, routine inulin analysis in feeds is uncommon [327]. It was not until very recently that the minimal performance requirements were established for fructans analysis in feed, pet food, and their ingredients [328]. However, excess dietary fructans have demonstrated adverse health effects in equines [329]. AOAC® (Rockville, Maryland, USA) Official Method^{SM} 997.08 is available to assess fructans in food products using ion exchange chromatography with pulsed amperometric detection. The method is based on two-step hydrolysis using amyloglucosidase (to remove starch) and inulinase. Measurement of simple sugars in different food-derived extract fractions is performed using glucoheptose as an internal standard. The same principle has been used to assess fructose derived from fructans in pet food [330]. Verspreet and coworkers analyzed fructan from wheat grains after acid hydrolysis. Mild conditions used during hydrolysis avoid the release from other naturally occurring saccharides in wheat that would otherwise interfere during the fructan estimation (e.g., raffinose oligosaccharides) [316]. Correia and coworkers have developed a method to analyze fructooligosaccharides [317]. These sugars are dietary and are used as food ingredients (incorporated as dietary fibers in commodities). The authors monitored sucrose pathway fermentation products from *Aspergillus aculeatus* as a potential source of fructooligosaccharides; fructose, glucose, sucrose, 1-kestose, nystose, and 1^F-Fructofuranosylnystose were monitored.

We have used enzymatic hydrolysis to obtain glucose from starch molecules present in feed and feed ingredients. Total and resistant starch was measured in several matrices including (e.g., silages) [318]. Bai and coworkers analyzed mono- and oligosaccharides from Hakka rice [319] as a measure of quality for sugars such as isomaltotriose, isomaltose, panose, maltose, and glucose. Finally, the determination of bacterial exopolysaccharides has also been reported [320,331].

4.3.7. Method Application Experience

Amine-based columns (e.g., Zorbax® Carbohydrate (Agilent technologies, Santa Clara, USA), Ultisil® XB-NH_2 (Welch Materials, Inc, Texas, USA)) are successful in separating mono and disaccharides in foods especially those containing lactose (such as dairy products). However, this type of stationary phase suffers easily from poisoning as amine functional groups form covalent bonding with several compounds (e.g., Schiff bases). As the amine functional group is sensitive to pH changes, extracts must be adjusted to avoid changes in the chemical form of the stationary phase functional group as this may affect repeatability/reproducibility or even obliterate the column capacity for retention. Hence, the elimination of interferences is paramount. Additionally, when retention capacity is lost, it is possible to apply changes in the mobile phase composition and flow (e.g., to increase acetonitrile concentration and reduce flow).

A particular case is that of coffee samples. Amine-based columns especially suffer when analyzing coffee extracts as they contain phenolic acids (e.g., chlorogenic, syringic, ferulic, protocatechuic and hydroxybenzoic acid) and alkaloids (e.g., caffeine, caffeic acid, theophylline, trigonelline) [332]. Costa Rican regulations accept not more than 10 g/100 g sucrose in roasted coffee. Hence, monitoring sugar levels, as a quality standard, in these products is paramount. When routine quality control in coffee samples is necessary, we recommend to use stationary phases more resistant to pH changes (e.g., amide-based), include mobile phase modifiers (e.g., triethylamine), or intensive extract clean up.

In the case of animal compound feed, for example, suckling pigs feed usually contain lactose. Contrary to the amine-based column (Figure 8A,B and Figure 9A), ion exclusion (e.g., Agilent Hi-Plex Ca, Phenomenex® Rezex™ RCU-USP Ca^{2+} (Torrance, California, USA)) is better equipped to deal with a larger range of samples and is less prone to deteriorate.

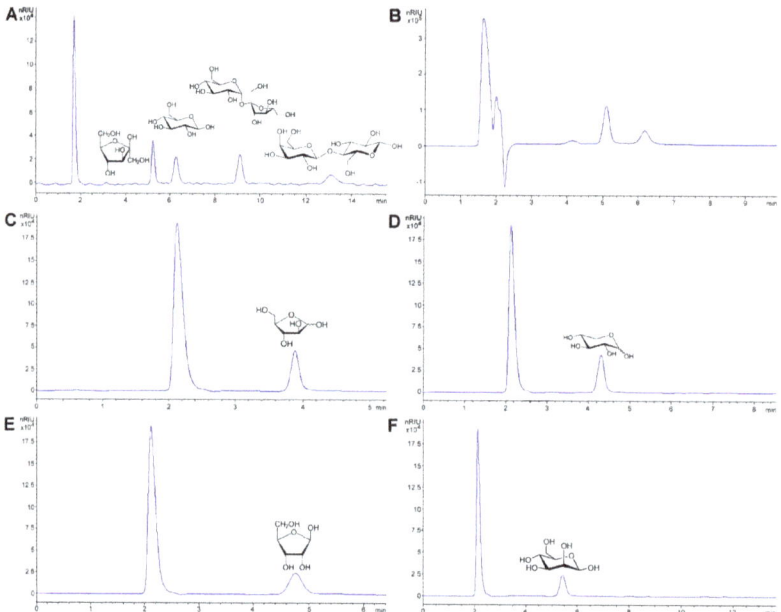

Figure 8. Chromatographs of (**A**) 2 g/100 mL standard mixture of four sugars including fructose (5.24 min), glucose (6.26 min), sucrose (9.12 min), and lactose (13.09 min) separated using amino column (Zorbax Carbohydrate, 0.7 mL min^{-1}, 80 ACN: 20 H$_2$O). (**B**) Sugar content of a molasses sample after hot water extraction, fructose (5.18 min) and glucose (6.31 min) signals are evident. (**C**) 1 g/100 mL standard solution for arabinose (3.89 min) (**D**) 1 g/100 mL standard solution for xylose (4.30 min) (**E**) 1 g/100 mL standard solution for ribose (4.76 min), and (**F**) 1 g/100 mL standard solution for mannose (5.42 min). Signal at ca. 1.80 min corresponds to the solvent front; constant in all injections.

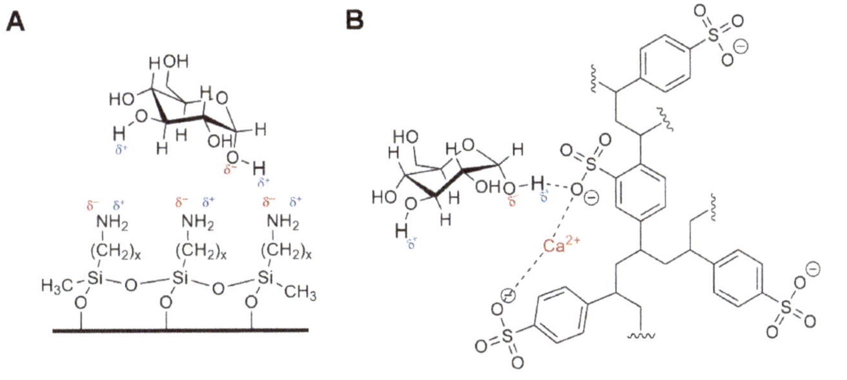

Figure 9. Schematic representation of sugar interaction mechanism using (**A**) amine based (**B**) calcium ion-based ligand exchange column.

These types of columns are able to separate H$^+$ (organic acids/monosaccharides), Na$^+$, Ca^{2+} (sugars/alcohols), Ag$^+$ and Pb^{2+} (oligosaccharides). Sugars and alcohols are separated using ligand exchange (Figure 9B) and organic acids by ion exchange. During ligand exchange, the more complex sugars elute first whereas simple ones, such as fructose, elute last, opposite to the elution order found in amine-based columns. Another advantage is that ion exchange columns need ultra-high purity water (type I) to segregate analytes while amino-based columns require acetonitrile in the mobile phase to perform. However, amino-based columns have the inherent advantage that complete chromatographic runs can be achieved under 12 min (while setting the column at 30 °C). Meanwhile, a good separation using exchange columns can be extended up to 30 min at 60 °C. As LC-MS usually use low flow rates to aid in solvent nebulization, it is harder to develop methods using this type of column due to their dimensions. When using ion exchange based-columns, EDTA can be used to sequestrate ions present on complex sample extracts reducing interferences. Peak tailing or fronting is also common when there is already wear of the chromatographic column. Finally, mathematic designs can be used to optimize method critical parameters and attributes, at least one research group has used this approach (i.e., Monte-Carlo simulation) to analyze sugars in herbs [333].

4.4. Organic Acids

Organic acids are common substances found naturally in several foods and result from fermentation processes [334]. As such, they are responsible for the particular flavor and aroma of commercially relevant commodities such as wine, vinegar, fermented meats, and yogurt, to name just a few. These substances have found widespread use in the food and feed industry as preservatives (increasing shelf-life, [335]) and antimicrobials (due to their bacteriostatic properties) (large-scale use of benzoic acid in beverages is a clear example, [335]). Organic acids can be determined by HPLC using ion exchange columns. Several anion exchange columns have already been mentioned in the previous section. Usually, though, measurement is commonly made using a UV detector at 210 nm (detection of absorption of carboxyl groups). Solvents and columns may vary slightly, but an isocratic method is sufficient to separate the analytes. The sample preparation for organic acid determination in beverages can be straightforward. In some cases, depending on the clarification, process suffered by the final product, it may consist of centrifugation and microfiltration. For an excellent primer for organic acids, we recommend the book edited by Vargas [336].

4.4.1. Reverse Phase Chromatography Analysis in Foods

Neffe-Skocińska and coworkers analyzed sugars, ethyl alcohol and organic acids in Kombucha tea beverages (a fermented brew) with an emphasis on glucuronic acid (which has been associated with health benefits) [337]. The fermentative profile was evaluated for 10 days at 3 different temperatures. Nour and coworkers use low temperature so they can separate 6 compounds (oxalic, tartaric, malic, lactic, citric, and ascorbic) in 13 min. The applied temperatures (i.e., 10 °C) in a 250 mm column using 0.7 mL min^{-1} flow rates ensures optimal resolution of the compounds while obtaining adequate peak shapes within a reasonable time. A buffer adjusted at 2.8 pH guarantees that the compounds of interest are maintained during chromatography as protonated species [338].

Different citrus juices were tested finding concentrations of citric acid ranging from 7.39×10^4 to 6.89×10^4 mg L^{-1}. Reverse phase separation of acids uses buffers (e.g., the $H_2PO_4^-/HPO_4^{2-}$ pair) or salts (e.g., Na_2SO_4) to accomplish separation. The advantage of this approach is that usually C_8/C_{18} columns are readily available and are relatively inexpensive. The downside resides in that the use of this kind of mobile phases increase the possibility of crystal precipitation in the pump and capillaries. The buffer has to be prepared daily (to circumvent microbial growth), and pH values strictly supervised (to avoid retention time shifts). Lobo Roriz and coworkers determined organic acids in three different medicinal plants which are widely consumed as infusions. *Gomphrena globosa* L. showed the highest levels or organic acids (mainly malic and oxalic) [339].

Pterospartum tridentatum (L.) Willk. and *Cymbopogon citratus* (DC.) Stapf showed higher levels of citric and succinic acids, respectively. Acid content depends on inherent plant genetic characteristics and edaphoclimatic conditions. The authors also analyzed sugars (using HPLC-RID Eurospher 100–5 NH$_2$ column and melezitose as internal standard) and, interestingly, tocopherols (α, γ, and δ-tocopherol, normal phase YMC Polyamide II column and fluorescence at λ_{ex} and λ_{em} 290 nm and 330 nm). Scherer and coworkers used a reverse phase column to assess ascorbic acid stability in apple, orange and lemon juices. They also compared nutritional analysis reported within the food labels for ascorbic acid with that obtained experimentally [340].

4.4.2. Ion Exchange Chromatography Analysis in Foods

Llano and coworkers analyzed sugars, acids, and furfural in pulp mill residue. The authors used a resin-based cross-linked gel column for low molecular-weight chain acids, alcohols, and furfurals. This method is particularly interesting since a comparison between 2 sets of columns for each application was tested (Table 15). The authors also included specific details for each column and optimize temperature, injection volume, and flow rate. Size exclusion also seem to have a role in sugar separation using ion exchange columns [341]. Though this application is not specifically for food, xylooligosaccharides (from revalorization alternatives for materials derived from the pulp mill enterprise) have found applications in the food industry and have been even linked to health benefits [342]. Saleh Zaky and coworkers reported a simultaneous analysis of chlorides, sugars, and acids [343]. However, the paper states that several inorganic ions are retained with the same strength within the column (all tested ions have different physicochemical properties; i.e., hydration spheres, charge among others). We have not been able to repeat this procedure. The authors did analyze sugars and acids (i.e., citric, lactic, acetic) and ethanol in a grand variety of food products including energy drinks, sodas, tomato juice and sauce, brine, milk, whey, cheese, and hummus. An interesting paper focused on the determination of organic acids from olive fruits. Different organic acid profiles were found for unique fruit varieties. They found oxalic, malic, succinic, and citric as main organic acids [344].

Table 15. Determination of organic acids and in foods and silage.

Matrix	Sample Pretreatment, Extraction	Mobile Phase Composition	Measurement Method, Chromatographic Column	Reference
		Reverse Phase-Based Columns		
Kombucha	H_2O 70–80 °C, filtration	20 mmol $H_2PO_4^-$ pH 2.4/MeOH (97:3)	Luna C_{18} 250 × 4.6 mm, 5 µm 30 °C UV λ 210 nm	[337]
Fresh fruits	Juice extraction, depulping, centrifugation	50 mmol L^{-1} $H_2PO_4^-$ pH 2.8	Hypersil Gold aQ 250 × 4.6 mm, 5 µm 10 °C UV λ 214 nm (254 ascorbic acid)	[338]
Medicinal plants infusions	$(HPO_3)_n$ extraction	3.6 mmol L^{-1} H_2SO_4	Sphere-Clone C_{18} 250 × 4.6 mm, 5 µm, 35 °C UV λ 215 nm (254 ascorbic acid)	[339]
Fruit juices	Filtration	0.01 mol L^{-1} KH_2PO_4 pH 2.6	RP-C_{18} 150 × 4.6 mm, 3 µm, UV λ 210 nm	[340]
Wine	SPE C_{18} Elution: mobile phase (for acid protonation)	0.005 mol L^{-1} H_3PO_4 pH 2.1, 1 mL/100 mL CAN	Lichrosorb RP-C_{18} 150 × 4.0 mm, 5 µm, UV λ 210 nm	[345]
		Ion Exchange-Based Columns		
Sulfite pulp mill	Filtration	1. Ultrapure water 79 °C 2. Ultrapure water 68 °C 3. H_2SO_4 0.005 mol L^{-1} 30 °C 4. H_2SO_4 0.005 mol L^{-1} 60 °C	RID for all cases, 1. Aminex HPX-87P Pb^{2+} 300 × 7.8 mm, 9 µm 2. Transgenomic® CHO-782 Pb^{2+} 300 × 7.8 mm, 7 µm 3. Bio-rad Aminex HPX-87H 300 × 7.8 mm, 9 µm 4. Shodex SH-1011 H^+ 300 × 8.0 mm, 6 µm	[341]
Food samples	Liquid samples: filtration Solid samples: H_2O 85 °C	0.005 mol L^{-1} H_2SO_4 35 °C	Hi-Plex H 300 × 8.0 mm, 6 µm, RID	[343]
Olive fruits	Maceration in H_2O MeOH (75:25)	0.1 g/100 mL H_3PO_4	Shodex RSpak KC-118 300 × 8.0 mm, UV λ 214 nm	[344]
Wines	Filtration, SPE strong anion exchange	0.065 mL/100 mL H_3PO_4	Aminex HPX-87H 300 × 7.8 mm, 9 µm, 65 °C, UV λ 210 nm	[346]
Fermented shrimp waste	Centrifugation, sonication, filtration	$(HPO_3)_n$ pH 2.1	SS Exil ODS 250 × 4.0 mm, 5 µm, UV λ 210 nm	[347]
Silage	H_2O 100 °C	6 mmol L^{-1} $HClO_4$	Shodex KC 811 300 × 8.0 mm, 7 µm 50 °C, UV λ 210	[348]
		Ion Exclusion-Based Analysis		
Drinks	Filtration, heat-aided degassing	Precondition: 10 mmol L^{-1} SDS 3 h 0.3 mL min^{-1} Elution: 1.84 mmol L^{-1} H_2SO_4 pH 2.43	Kinetex XB-C_{18} 150 × 4.6 mm, 2.6 µm, UPLC-UV λ 210 nm	[349]

Mihaljević and coworkers separated organic acids in wine. Organic acid profile (especially glucuronic and galacturonic acids levels) was able to distinguish among Traminer vs. Welsch produced Croatian wines. Mobile phase rate was reduced during chromatography when target acids were glucuronic, gluconic, galactaric, and galacturonic [346]. Diacids and citric acid considerably differ structurally (e.g., number of carbons). Meanwhile, the reduction of flow rate responds to the subtle differences among these intimately related structures, making them more difficult to resolve. Sánchez-Machado and coworkers preserved shrimp tissue through fermentation with lactic acid bacteria. A complete separation of lactic, citric and acetic acid was accomplished. Sonication time and initial sample mass were optimized during the assay [347]. Finally, though most of the tests regarding organic acids extraction-wise are straightforward, even in brightly colored samples (e.g., fruits [350]), still SPE cleanup has been applied, with adequate recoveries, to these extracts to remove interferences as anthocyanins and carbohydrates that may co-elute during acid analysis (especially relevant if a non-selective detector is used) [345].

4.4.3. Ion Exclusion Chromatography Analysis in Foods

Fasciano and coworkers modified a reverse phase column, a C_{18} column was dynamically modified by running a solution of SDS through the column (Table 15). They separated organic acids after optimizing sulfuric acid concentration, flow rate, and pH; an example of how a reverse column can be made more versatile. A wide array of compounds in juices and sodas were analyzed using ion exclusion chromatography [349]. For a detailed description of ion exclusion chromatography, we encourage the reader to pay special attention to this paper introduction.

4.4.4. Silages

The maturity of the crop governs silage quality at harvest. However, fermentation in the silo further influences the nutritive value of silage. Coblentz and Akins recently published a detailed discussion of silages [351]. Similarly, Khan and coworkers wrote a more specific review based on maize silages [352]. In both papers, references to silage quality based on organic acids are mentioned. Since silage is the result of this fermentation process, the organic acid analysis is used to monitor its quality. Concentrations of fermentation acids do not seem closely related to silage intake; however, they are decisive in the balance of volatile fatty acids produced in the rumen. In turn, affecting gluconeogenic metabolism and influencing milk and body composition in productive livestock. Several researchers have dedicated efforts to not only assess organic acid concentrations from silages but also have studied the effect that organic acid has on silage fermentation. For example, Ke and coworkers included malic or citric acid at concentrations of 0.1 to 0.5 g/100 g during alfalfa ensiling of alfalfa and concluded that these levels improved silage fermentation quality [348]. Additionally, both acids can be further used as feed additives that have proven to promote animal performance. Silva and coworkers determined the fermentation profile of alfalfa silages treated with microbial inoculants at different fermentation periods under tropical conditions [353]. From the strains tested *P. pentosaceus* showed the most efficiency suggesting its use as a silage inoculant. The sample pretreatment just consisted of extract acidification with metaphosphoric acid, gravity-aided filtration, and centrifugation.

4.4.5. Method Application Experience

We have used ligand exchange-based analysis to routinely screen silage quality (Figure 10). Sample pretreatment consists of metaphosphoric acid extraction. We also have taken advantage of sample extraction for ammoniacal nitrogen (a modified version of method 941.04). In this type of columns, poly and diacids are eluted first. Monocarboxylic acids will elute later on during the chromatographic run in order of increasing alkyl chain length (i.e., formic, acetic, propionic, butyric). Finally, we have used liquid chromatography coupled with a variable wavelength detector set at 210 nm, a Hi-Plex H (300 × 7.7 mm and 8 μm particle size) column kept at 60 °C, and a 50 mmol L^{-1} H$_2$SO$_4$ solution with a flow rate of 0.6 mL min^{-1} to monitor ammonium propionate, added as a preservative, in dry dog foods (see for example, [354]). Average concentrations of (693.12 ± 75.63) mg kg^{-1} have been obtained for local products. Sieved (at 0.5 mm particle size) dog food was treated with hot water to extract the propionate quantitatively. Both RID and UV detectors can be used for both organic acid and sugar (and alcohol) analysis. Using RID will enable the user to monitor all the compounds above simultaneously, but RID detectors suffer from low sensitivity when compared to others.

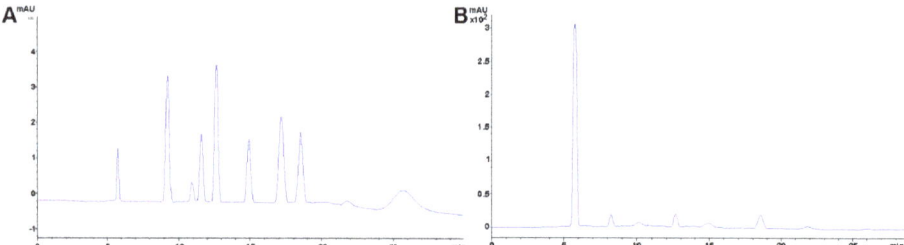

Figure 10. Chromatographs of (**A**) Mix of organic acid standards malic acid (9.24 min) methanoic acid (formic acid, 10.92 min), ethanoic acid (acetic acid, 11.65 min), propanoic acid (propionic acid, 12.62 min), lactic acid (14.92 min), 2-methylpropanoic acid (isobutyric acid, 17.22 min), butanoic acid (butyric acid, 18.52 min). (**B**) A silage sample after extraction with acid 0.01 mol L^{-1} H$_2$SO$_4$. Fermentation products identified at 18.499 min, 14.903 min, 12.606 min. The signal at ca. 5.70 min corresponds to the solvent front.

4.5. Vitamins

Vitamins are essential micronutrients that humans and animals need for normal metabolism. The lack of these nutrients in dietary sources can cause serious disease; trace amounts of these compounds are required for growth and reproduction. Based on their solubility, vitamins have been divided into two groups: those soluble in organic non-polar solvents and water-soluble vitamins. Thus, the vitamins from the B-complex and vitamin C are classified water soluble while the fat-soluble vitamins are isoprenoid compounds, namely vitamins A, D, E and K. This last group is found in small amounts on foodstuffs, associated with lipids; stored in the liver and fatty tissues, and are eliminated slower than water-soluble vitamins [45].

4.5.1. Fat-Soluble Vitamins

Vitamin A is commonly expressed as retinol equivalents but can occur in different chemical forms, i.e., retinal, retinoic acid and retinyl esters. In foods, it is very common to find this vitamin as retinyl esters, more specifically, as acetate, propionate, or palmitate [355]. This vitamin is involved in immune function, vision, reproduction, and cellular communication [356]. Vitamin E is a term used to designate some related compounds as tocopherols and tocotrienols, it is found in fat products of vegetal origin, mainly oils. The most common tocopherols that can be found in food and feed are $\alpha/\beta/\delta/\gamma$-tocopherol in different proportion. Mixed tocopherols are considered the most effective lipid-soluble antioxidants [357]. Vitamin D is naturally present in very few foods like sea products, eggs, meat, and dairy products, the most commonly found members are vitamin D_2 and D_3, but it is also produced endogenously when ultraviolet rays from sunlight strike the skin and trigger vitamin D synthesis from 7-dehydrocholesterol [358]. It promotes the absorption of calcium, regulates bone growth and plays a role in immune function [359]. Lastly, vitamin K is an essential nutrient for animals and humans because it is required for functioning of the blood clotting cascade [360], just as vitamin D, vitamin K can be found in two forms i.e., phylloquinone (vitamin K_1, found in green plant leaves e.g., spinach, collards, lettuce, and broccoli [361]) and menaquinone (vitamin K_2, bacterium residing in the vertebrate intestine [362]).

Since these compounds are involved in metabolic pathways, and are paramount in health promotion in animals and humans, it is crucial to determine their content in food and feed to comply with daily requirements and quality control. That is why several studies have been conducted regarding the extraction and quantitative analysis of these vitamins, either individually or simultaneously [359,363–382].

Sample Preparation

Most of the analytical methods involve previous steps of sample preparation like saponification, solid-liquid or liquid–liquid extractions, followed by a concentration step before HPLC analysis (Table 16). The sample pre-treatment is critical for an accurate method. That is why there are many aspects that need to be controlled, Qian and Sheng have studied seven different variables to take into account for simultaneous analysis of vitamins in animal feed. These variables were related to the extraction procedure: (1) sample particle size, (2) solvent, (3) the ratio of sample to solvent, (4) extraction with and without N_2 protection, (5) extraction time, (6) equipment and (7) the use of SPE for cleanup [367]. They evaluated how each of the variables affected both the coefficient of variation and the recovery of each of the vitamins in order to obtain extraction conditions that would allow them to satisfy each of the vitamins in a satisfying way.

Table 16. Determination of fat vitamins and in foods and feeds.

Matrix	Vitamin	Extraction Method	Measurement Method, Chromatographic Column	Reference
Infant and nutritional formulas	K_1	Lipase at 37 °C 2 h in PBS Extraction with hexane and concentrate with N_2, reconstituted with MeOH	LC-MS/MS-ESI$^+$ C_{18} 50 × 2.1 mm, 2.6 µm 40 °C. Mobile phase: H_2O/ACN 50:50 with 0.1 mL/100 mL HCOOH and ACN/MeOH 75:25, with 2.5 mmol L^{-1} NH_4CO_2H, gradient system	[377]
Additives, premixes, complete feed	A, E, K_3, D_3	Extraction with 0.2 g/100 mL NH_3 and EtOH, sonication 40–50 °C for 20–30 min in the dark, SPE (OASIS HLB)	HPLC-PDA C_{18} 150 × 4.6 mm, 5 µm. Mobile phase MeOH/H_2O (98:2), UV λ 230 nm	[366]
Okra (Abelmoschus esculentus)	B_2, B_3, B_6, B_{12}, C, E, K, D, A, β-Carotene	Vitamin B Analyses. 0.1 mol L^{-1} H_2SO_4 incubation 30 min at 121 °C, pH 4.5, 2.5 mol L^{-1} $NaC_2H_3O_2$, Takadiastase enzyme Ascorbic Acid Analyses Extraction 0.3 mol L^{-1} metaphosphoric acid and 1.4 mol L^{-1} HAoc Saponification KOH 50 g/10 g in EtOH in reflux at 50 °C for 40 min. Extraction with ether	Vitamin B Analyses. Agilent ZORBAX Eclipse Plus C_{18} 250 × 4.6 mm, 5 µm. Mobile phase MeOH/0.023 mol L^{-1} H_3PO_4 pH 3.54 (33:67) UV λ 270 nm at room temperature Ascorbic Acid Analyses PDA, 0.1 mol L^{-1} $KC_2H_3O_2$ pH 4.9 and ACN/H_2O (33:67). UV λ 254 nm at room temperature β-carotene: Agilent TC-C_{18} 250 × 4.6 mm, 5 µm, ACN/MeOH/ethyl acetate (88:10:2) at 453 nm. Fat-soluble vitamins Eclipse XDB-C_{18} 150 × 4.6 mm, 5 µm, MeOH at: 325, 265, 290, 244 nm	[363]
Fruits, juices, and supplements	A, D_3, E, B_1, B_2, B_3, B_5, B_6, B_9, B_{12} C	Extraction with Carrez I and Carrez II solutions Cleaned by SPE (RP$_{18}$ Bakerbond)	PDA, TSKGel ODS-100V, 150 × 4.6 mm, 5 µm at 30 °C. Gradient 0.01 mL/100 mL TFA in H_2O and MeOH at 320, 275, 253, 290, 258, 218, 289, 360, and 262 nm	[364]
Cheeses	D_3	Saponification with KOH (60% g/100 mL in H_2O). Extraction: MeOH/$CHCl_3$	DAD, at 228 and 266. C_{18} 200 × 4.6 mm, 5 µm ambient temperature Mobile phase: MeOH/ACN/H_2O (49.5:49.5:1)	[378]
Dairy and soybean oil	K_1, D_3, E, A	Saponification: KOH/EtOH + ascorbic acid. Mechanical shaking. Extraction: hexane, evaporated at 40 °C and re-dissolved in MeOH	PDA, C_{18} 250 × 4.6 mm, 5 µm at 30 °C, 3.0 g/100 mL SDS and 0.02 mol L^{-1} PBS at pH 7, with 15.0 mL/100 mL butyl alcohol (organic solvent modifier). 230 and 300 nm (K_1, D_3 and E vitamins), 280 nm (A, E, D_3, and K_1)	[379]
Milk sample	K_1, D_3, D_2, E, A	Extraction: hexane, concentrate with N_2 and reconstituted with MeOH	HPLC-UV-Vis 325, 264, and 280 nm (A, D, K and E), C_{18} DBS 150 × 3 mm, 3 µm at room temperature; MeOH/H_2O (99:1) and MeOH/THF (70:30)	[380]
Vegetables, fruits, and berries	K_1	Extraction with 2-propanol/hexane and $CHCl_3$/MeOH	HPLC-EC Vydac 201 TP54 150 × 4.6 mm, 5 µm. Mobile phase: MeOH/0.05 mol L^{-1} $NaC_2H_3O_2$, pH 3 (96:4)	[381]
Rice	E	Saponification KOH (600 g L^{-1}) in EtOH with pyrogallol for 45 min at 70 °C, mechanical shaking. Microextraction: CCl_4/ACN (1:10)	FLD, COSMOSIL π-NAP 250 × 4.6 mm, 5 µm at 25 °C. Mobile phase: MeOH/H_2O/ACN (80:13:7). Vitamin E isomers λ_{ex} 290 and λ_{em} 330 nm	[382]

Regarding the first variable in meals and flours, subsample variability and homogeneity is closely linked to particle size. Qian and Sheng observed that large sample particle size causes an incomplete vitamin A extraction with high variability [367]. We have noted that for fresh products, with a total fat content greater than 10 g/100 g and high moisture content (i.e., greater than 85%) (e.g., avocado (*Persea americana* Mill.) and peach palm (*Bactris gasipaes* Kunth)), it is advisable to freeze dry the sample, before analysis, to promote homogenization and eliminate water that interferes with fat-soluble compound extraction.

Qian and Sheng developed the assay procedure without saponification, but nevertheless it is a widespread procedure used in the analysis of vitamins since it is an efficient way of removing interferences of lipid origin that can be found in the matrix [363,365,368,374,378,379,382]. Since it is based on an alkaline digestion (i.e., heated KOH or NaOH aqueous or alcoholic solutions) there is a disadvantage, the saponification could generate oxidation of the vitamins, which translate into a loss for vitamin degradation and low recovery percentage [366,368]. Some researchers have made use of the antioxidant ability of some compounds such as BHT, BHA, TBHQ, ascorbic acid or pyrogallol to reduce oxidation losses [363,365,367,368,371,374,379,380]. Nevertheless, saponification procedures take time, and the extractions procedures are not always straightforward, because emulsions are generated, as Lim and co-worker mentioned in their comparison of extraction methods for determining tocopherols in soybeans [369], these inconveniences introducing considerable variation, low recovery, and reproducibility, situations that we have also seen in the development of this type of methodologies.

Recently, alternatives to saponification process for the extraction of vitamins have been used, some papers use enzyme-catalyzed hydrolysis and alcoholysis of ester bonds in vitamin A and E esters to facilitate their determination in milk powder and infant formula. They assayed six lipase preparations and one esterase preparation using diisopropyl ether, hexanes/ethanol and supercritical CO_2 containing ethanol. Three of the lipases' preparations from *Candida antarctica* (Novozyme 435), *Rhizomucor miehei* (Lipozyme IM) and *Pseudomonas cepacia*, showed considerably higher activity toward retinyl palmitate but there was no observed activity with α-tocopheryl acetate [372]. In feed, Xue and co-workers applied enzymolysis instead of saponification with a basic proteinase named Savinase in 30 min of incubation time at 40 °C getting good results in the determination of four fat-soluble vitamins (K_3, A, D_3, E) [366].

With respect to the solvent type and ratio solvent: sample, there is a wide variety of solvents available for the fat-soluble extraction, most of the methods use solvents such as hexane, heptane, chloroform, dichloromethane, ethyl acetate, tetrahydrofuran, ethyl ether, and the choice will depend on the type of matrix to work with (Table 16). For example, in the case of animal feed [367], a poor resolution was observed using hexane and chloroform, generating an overestimation of vitamin D, such mixture does not allow a good separation during centrifugation which produced a high %RSD.

If a mixture of acetone/$CHCl_3$ (30:70) is used, the results in terms of variability and recovery of vitamins are outstanding, mainly for vitamin A. In low-fat matrices (less than 0.1 g/100 g, e.g., fruit juices), this solvent system has the disadvantage of generating emulsions and, hence, low recoveries. In the case of dairy and infant formulas where the presence of milk proteins is a hindrance, the extraction of the lipid part has been reported using saponification and extraction with hexane, leading to vitamin degradation in fat. It is also an extensive process [373,374]. For this reason, a group of researchers developed a fat extraction methodology using a mixture of CH_2Cl_2:EtOH 2:1 and separation at 4 °C with a centrifuge, giving satisfactory results for analysis of FAMES so it could be applied in the extraction of vitamins in these matrices [375].

Regarding extraction time and equipment, Qian and Sheng used vortex mixer for several minutes, rotatory mixer and supersonic mixer, these last two methods were not as effective for extraction of vitamin A and other vitamins due to low recoveries [367]. Hung used a rotatory mixer for extraction of vitamins D_2 and D_3 during one hour [376]. We have found, that for foodstuffs, the most efficient sample treatment is to rely on the combination of a vortex mixer for one minute, a rotatory mixer for 30 min or supersonic mixer for 15 min.

Effective extraction can be aided if the solvent contains a percentage of an appropriate antioxidant. N_2 has been used in some protocols to the protection [377,380], of extracted vitamins from degradation because the solvent vapor that replaces air over the surface of extraction mixture has a protective antioxidant effect [367]. Qian and Sheng showed evidence that this protection did not influence in the mean values of vitamins A, D, and E and pro-vitamin D, but decreased the variation coefficient [367].

Chromatographic Analysis

The analytical method for the determination of vitamins in food and feed, is liquid chromatography (HPLC or UPLC) due to its, selectivity, short time of analysis, and high resolution. Methods based on chromatography can be easily automated and can determine several compounds at the same time.

Methods range from using normal phase chromatography with silica columns to reverse phase chromatography with C_8, C_{18}, and C_{30} columns (Table 16). Lee and coworkers studied three different columns to separate vitamin A and E: an NH_2 column, C_{30}, and C_{18}. Concerning the resolution, they observed the β-tocopherol and γ-tocopherol peaks of vitamin E were not separated and appeared as a single overlapping peak when using a C_{18}, but it could be separated using an NH_2 column. Regarding detection and quantification limits the NH_2 column presented values lower than C_8 column but higher than C_{18}.

The solvent systems to use as mobile phase vary depending on the selected approach, in the case of normal phase chromatography the solvents systems mostly used are 2-propanol/hexane in different proportion, but also can be use methanol/hexane/THF (97.25:2.5:0.25), or hexane/MTBE (96:4) [374]. In reverse phase the most common are MeOH-H_2O, MeOH-ACN, both techniques can be used in gradient o isocratic mode.

As mentioned before, the liquid chromatography technique has a wide variety of monitoring techniques including PDA, FLD, ECD, ELSD or MSD. The most commonly used detector for vitamins is FLD, which is considerably more sensitive and selective than UV. Therefore, it is possible to carry out a simultaneous determination of vitamin A and E, for which a programming of the equipment is required so that at certain time intervals it uses the excitation wavelength (λ_{ex}) and emission wavelength (λ_{em}) specifies for each vitamin, for vitamin E, λ_{ex} = 285 and λ_{em} = 310 nm, for vitamin A the configuration at λ_{ex} = 325 and λ_{em} = 470 nm, but no other vitamins could be detected such as K or D. Alternatively, PDA can work with multiple UV wavelengths and determine the four vitamins at the same time. Mass spectrometry coupled chromatography is usually the most versatile option. However, it requires that the laboratory has the resources for its acquisition. We have successfully applied mass spectrometry to assess tocopherols in feed supplements and animal biological samples (Figure 11A–F).

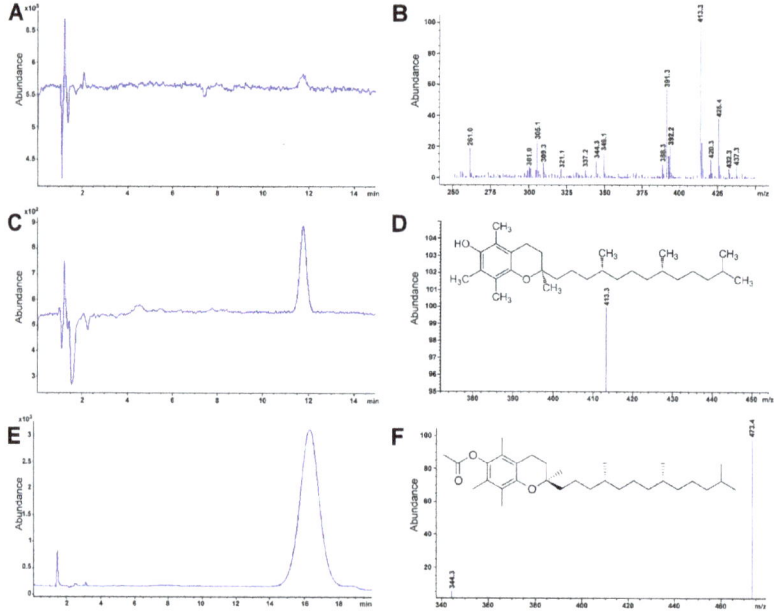

Figure 11. Single quadrupole LC/MS ESI$^+$ chromatographs of (**A**) Total ion chromatogram α-tocopherol (a 1 mg L^{-1} solution in butanol) signal positively identified at 11.75 min (**B**) Mass spectra for α-tocopherol (a 1 mg L^{-1} solution in butanol) using a cone energy of 120 V extracted from a signal with a retention time of 11.71 min (**C**) α-tocopherol (retention time 11.77 min) identified in a chicken plasma sample after extraction with chloroform and butanol (**D**) α-tocopherol in selected ion monitoring (SIM) mode using a cone energy of 120 V extracted from signal with a retention time of 11.82 min (**E**). α-tocopherol acetate in an injectable vitamin E solution for veterinary use using a "dilute and shoot" approach (16.32 min), and (**F**) α-tocopherol acetate in SIM mode using a cone energy of 60 V extracted from signal with a retention time of 16.34 min.

4.5.2. Hydrosoluble Vitamins

One of the main issues that the hydrosoluble vitamins analysis exhibit is that each molecule is structurally different. Hence, to assess each vitamin, different conditions must be applied to the HPLC system to assess each vitamin. Kim published a paper in which ion pairing chromatography was used to monitor six different vitamins (nicotinic acid, nicotinamide, folic acid, and pyridoxine) in the feed [383]. Sodium hexanosulfate was used to with this approach; all six vitamins can be quantified using the same chromatographic run (using the same wavelength and column for all species) (Figure 12).

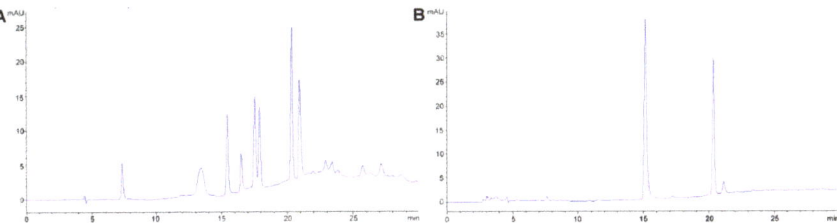

Figure 12. Hydrosoluble vitamin analysis based on ion pairing [383]. (**A**) Successful separation of 7 complex B vitamins including niacin (nicotinic acid, B_3, 6.67 min), FMN (B_2, 14.12 min), pyridoxal (B_6, 17.007 min), pyridoxamine (B_6, 18.607 min), pyridoxine (B6, 19.963 min), folic acid (B9, 20.630 min), and thiamine (B_1, 25.074 min). (**B**) Analysis of a vitamin premix destined for feed formulation. Another advantage presented is that the separation can be performed using a reverse phase C_{18} column.

A usual problem that arises with the analysis of this compounds is the fact that pH changes affect their chemical behavior drastically both during extraction and chromatographic separation. The author circumvented this issue using a mobile phase and sample extraction solution spiked with 0.1 mL/100 mL acetic acid, maintaining all species protonated. Midttun and coworkers described an extensive two-phase analysis based on an LC-MS/MS (to determine fat soluble and water soluble). In the chloroform/isooctane phase all-*trans* retinol, 25-hydroxyvitamin D_2, 25-hydroxyvitamin D_3, α-tocopherol, γ-tocopherol, and phylloquinone were retained. The hydrophilic phase (in which water-soluble vitamins were found), was mixed with ethanol, water, pyridine, and methyl chloroformate as a derivatizing agent. In this assay there can be a third phase (i.e., the methyl chloroformate fraction) that it is reserved for gas chromatography analysis of amino acids [384]. As an excellent example of hydrosoluble vitamins analysis using LC-MS/MS in the food industry, is the determination of 15 compounds in beverages using a multi-mode column (SM-C_{18} column, 150 × 2.0 mm, 3 µm; Imtakt Co., Kyoto, Japan), which provided reverse-phase, anion- and cation-exchange capacities, and therefore improved the retention of highly polar analytes such as water-soluble vitamins. The use of this column removes the need for an ion pair reagent in the mobile phase [385]. Finally, we encourage the reader toward a recent and ample review regarding fat- and hydrosoluble vitamins, respectively [386,387].

4.5.3. Method Application Experience

In our experience in the development of a methodology for the determination of fat-soluble vitamins in food matrices, the most challenging part has been the sample pretreatment. As mentioned above, several factors have to be considered. For the species retinyl acetate and palmitate, we chose to use a direct extraction to avoid decomposition by the saponification process. Extraction can be performed with hexane, ethyl acetate or chloroform; as the last solvent is far easier to eliminate, during concentration steps, is considered the most suitable option. Isopropanol is a useful solvent for reconstitution.

This method applies to dry matrices as flour, bakery products, freeze-dried pulps or fortified sugar. In the case of dairy products, we highly recommend the use of dichloromethane/ethanol. Separation is carried out using an HPLC-DAD set at 325 nm and a Zorbax Eclipse XDB-C_8 (150 × 4.6 mm, 5 µm) column at 50 °C. Shifting the solvent system form a MeOH/H_2O (90:10) to MeOH/2-propanol/ACN (95:1.5:3.5) saves up to 5 min of chromatographic run time and better peak shape, for the palmitate, is obtained (Figure 13A,B).

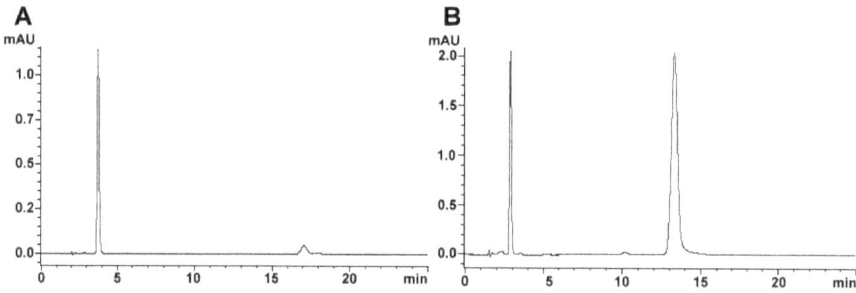

Figure 13. Chromatographs for vitamin A standards mixtures separated with C_8 column at 325 nm and 50 °C, of (**A**) retinyl acetate (3.11 min) and retinyl palmitate (17.63 min) using MeOH/H_2O (90:10) and (**B**) retinyl acetate (2.89 min) and retinyl palmitate (13.30 min) using MeOH/2-propanol/acetonitrile (95:1.5:3.5).

Monitoring the analytes at 295 nm, using MeOH/H_2O (90:10) as solvents at a flow rate of 0.5 mL min^{-1}, tocopherols can be successfully separated (i.e., retention times for δ/γ/α-tocopherol 5.35, 6.03 and 8.59 min, respectively; R_s 1.97). In samples with relatively high-fat content, saponification is necessary to eliminate interferences that usually share similar retention times that those for α-tocopherol. Food and feed samples can be fortified or can naturally contain both, vitamin D_2 and D_3. Therefore, for a method to be suitable, simultaneous detection of both analytes is a must. Under the conditions above, C_8 stationary phases are incapable of resolving both species.

A C_{18} column heated at 30 °C and a MeOH/2-propanol/ACN (90:3:7) solvent system, with a flow set at 0.3 mL min^{-1}, can achieve a resolution of 1.25 (Figure 14A,B). Though a mobile phase composed MeOH and H_2O is highly desirable, a drawback using this solvent system in complex matrices is that α-tocopherol can be interference for the identification of vitamin D_3 and vice versa (Figure 14C). A solvent gradient, a column with longer alkyl chains (e.g., C_{30}) or the use an MS detector may be employed to solve this issue.

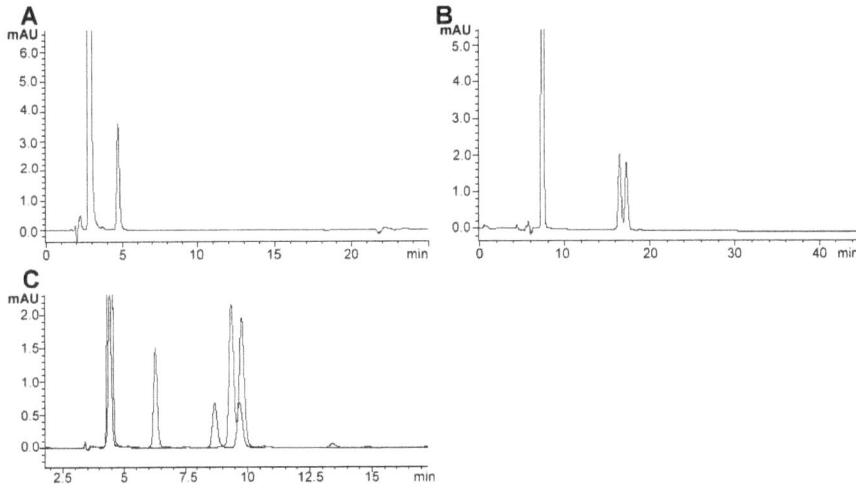

Figure 14. Separation for vitamin D_2+D_3 standards at 264 nm using (**A**) a C_8 column and (**B**) a C_{18} column D_2 (16.47 min) y D_3 (17.24 min). Analysis performed at 30 °C using MeOH/2-propanol/ACN (90:3:7) (**C**) Superposed chromatograms for vitamin $D_2 + D_3$ (blue line) and $\delta/\gamma/\alpha$-tocopherol standards using a C_{18} column and MeOH/H_2O (90:10), 30 °C.

5. Conclusions

LC is a powerful and versatile tool for food and feeds analysis, food and feed matrices are complex mixtures that on occasion present to the researcher difficulties as analytes of interest must be extracted and purified before injecting into the LC system. Advantages that chromatography provides when applied to food or feed analysis include sensitivity (determination of trace amounts, especially important in the case of contaminants, residues, controlled or undesired substances), automation and high throughput (reducing time and user dependence in laboratories with considerable workloads), simultaneous determination of multiple analytes. Food and feed chemists must make an effort to develop methods that provide a faster response and with few possible numbers of steps. Several current methods that are based on other step-full or less automated, or specific techniques can be reinvented, transformed and transplanted to LC analysis to improve sensitivity, specificity, and selectivity (For example, exchanging a spectrophotometric-based piperine analysis for a chromatography one). Within the myriad of alternatives, the LC approach delivers, each on its own seldom can solve a research problem, and each technique (i.e., each detector, each chromatographic column or sample treatment) has its shortcomings and limits as to which data is to provide. Usually, a multiphasic methodology is desirable to reach an appropriate conclusion. Hence, LC methods are more useful when are tailored to fit a purpose. Nowadays, mass spectrometry coupled liquid chromatography is an almost widespread technique that can provide molecule confirmation, ease trace analysis and allows the assay of multiple analytes simultaneously. Not only is a versatile tool for routine analysis but, research-wise, it provides more information about the target molecules and opens a valuable doorway toward a myriad of applications in food analysis including metabolomics, proteomics, and parvomics. Notwithstanding, as demonstrated above, traditional detectors are still the most commonly available in most laboratories. With a proper sample pretreatment, some traditional-detector-based methods are, regarding analytical performance, comparable to those based on MS.

Author Contributions: Conceptualization: C.C.-H and F.G.-C.; Methodology: C.C.-H, G.A., F.G.-C., A.L.; Software: F.G.-C.; Investigation: C.C.-H, G.A., F.G.-C., A.L.; Resources: C.C.-H, G.A., F.G.-C., A.L.; Data Curation: F.G.-C.; Writing—original draft preparation: C.C.-H, G.A., F.G.-C., A.L., C.C.-H, G.A., F.G.-C., A.L.; Visualization: F.G.-C.; Supervision: F.G.-C, G.A; Funding Acquisition: C.C.-H.

Funding: This research was funded by the University of Costa Rica grants number A2502, B2062, B2066, B2659, B8042, B5084. B3097, ED-427 and ED-428 and the APC was funded by the Office of the Vice Provost for Research of the University of Costa Rica.

Acknowledgments: Marelyn Rojas Lezama is acknowledged for her help doing the experiments regarding nitrate and nitrite in the hay. Special thanks to Guy Lamoureux Lamontagne for his suggestions.

Conflicts of Interest: The authors declare no conflict of interest.

References

1. Ercsey-Ravasz, M.; Toroczkai, Z.; Lakner, Z.; Baranyu, J. Complexity of the international agro-food trade network and its impact on food safety. *PLoS ONE* **2012**, *7*, e37810. [CrossRef]
2. Fair, K.R.; Bauch, C.T.; Anand, M. Dynamics of the Global Wheat Trade Network and Resilience to Shocks. *Sci. Rep.* **2017**, *7*, 7177. [CrossRef] [PubMed]
3. Keding, G.B.; Schneider, K.; Jordan, I. Production and processing of foods as core aspects of nutrition-sensitive agriculture and sustainable diets. *Food Secur.* **2013**, *5*, 825–846. [CrossRef]
4. Canady, R.; Lane, R.; Paoli, G.; Wilson, M.; Bialk, H.; Hermansky, S.; Kobielush, B.; Li, J.-E.; Llewellyn, C.; Scimeca, J. Determining the applicability of threshold of toxicological concern approaches to substances found in foods. *Crit. Rev. Food Sci. Nutr.* **2013**, *53*, 1239–1249. [CrossRef] [PubMed]
5. Van der Fels-Klerx, H.J.; Adamse, P.; de Jong, J.; Hoogenboom, R.; de Nijs, M.; Bikker, P. A model for risk-based monitoring of contaminants in feed ingredients. *Food Control* **2017**, *72*, 211–218. [CrossRef]
6. Allard, D.G. The 'farm to plate' approach to food safety—Everyone's business. *Can. J. Infect. Dis.* **2002**, *13*, 185–190. [CrossRef]
7. Kniel, K.E.; Kumar, D.; Thakur, S. Understanding the complexities of food safety using a "One Health" approach. *Microbiol. Spectr.* **2018**, *6*. [CrossRef]
8. Krska, R.; de Nijs, M.; McNerney, O.; Pichler, M.; Gilbert, J.; Edwards, S.; Suman, M.; Magan, N.; Rossi, V.; van der Fels-Klerx, H.J.; et al. Safe food and feed through an integrated toolbox for mycotoxin management: The MyToolBox approach. *World Mycotoxin J.* **2016**, *9*, 487–495. [CrossRef]
9. Wartella, W.A.; Lichtenstein, A.H.; Boon, C.S. *Front-of-Package Nutrition Rating Systems and Symbols*; Institute of Medicine of the National Academies Press: Washington, DC, USA, 2010; ISBN 978-0-309-15827-5.
10. Association of American Feed Control Officials (AAFCO). Pet feed regulation. In *AAFCO Official Publication*; AAFCO: Atlanta, GA, USA, 2018.
11. Yashin, Y.I.; Yashin, A.Y. Analysis of food products and beverages using high-performance liquid chromatography and ion chromatography with electrochemical detectors. *J. Anal. Chem.* **2004**, *59*, 1237–1243. [CrossRef]
12. Nollet, L.M.L. Liquid chromatography in food analysis. In *Encyclopedia of Analytical Chemistry*; Meyers, R.A., McGorrin, R.J., Eds.; John Wiley and Sons: Hoboken, NJ, USA, 2006. [CrossRef]
13. Quitmann, H.; Fan, R.; Czermak, P. Acidic organic compounds in beverage, food, and feed production. In *Biotechnology of Food and Feed Additives*; Zorn, H., Czermak, P., Eds.; Springer: Berlin/Heidelberg, Germany, 2014; ISBN 978-3-662-43760-5.
14. Jiang, J.; Xiong, Y.L. Natural antioxidants as food and feed additives to promote health benefits and quality of meat products: A review. *Meat Sci.* **2016**, *120*, 107–117. [CrossRef] [PubMed]
15. Candan, T.; Bağdatl, A. Use of natural antioxidants in poultry meat. *CBU J. Sci.* **2017**, *13*, 279–291.
16. Chanadang, S.; Koppel, K.; Aldrich, G. The impact of rendered protein meal oxidation level on shelf-life, sensory characteristics, and acceptability in extruded pet food. *Animals* **2016**, *6*, 44. [CrossRef] [PubMed]
17. Sanches-Silva, A.; Costa, D.; Alburquerque, T.G.; Buonocore, G.G.; Ramos, F.; Castilho, M.C.; Machado, A.V.; Costa, H.S. Trends in the use of natural antioxidants in active food packaging: A review. *Food Addit. Contam. Part A* **2014**, *31*, 374–395. [CrossRef] [PubMed]

18. Erickson, M.C.; Doyle, M.P. The challenges of eliminating or substituting antimicrobial preservatives in foods. *Annu. Rev. Food Sci. Technol.* **2017**, *8*, 371–390. [CrossRef]
19. Gupta, C.; Prakash, D.; Gupta, S. A biotechnological approach to microbial based perfumes and flavours. *J. Microbiol. Exp.* **2015**, *2*. [CrossRef]
20. Das, A.; Chakraborty, R. An introduction to sweeteners. In *Sweeteners, Reference Series in Phytochemistry*; Mérillon, J.-M., Ramawat, K.G., Eds.; Springer International Publishing: Basel, Switzerland, 2018; ISBN 978-3-319-27026-5.
21. Mooradian, A.D.; Smith, M.; Tokuda, M. The role of artificial and natural sweeteners in reducing the consumption of table sugar: A narrative review. *Clin. Nutr. ESPEN* **2017**, *18*, 1–8. [CrossRef]
22. Durán, S.; Dávila, L.A.; Contreras, M.C.E.; Rojas, D.; Costa, J. Noncaloric sweeteners in children: A controversial theme. *BioMed Res. Int.* **2018**, *2018*, 4806534.
23. Subedi, B.; Kannan, K. Fate of Artificial Sweeteners in Wastewater Treatment Plants in New York State, USA. *Environ. Sci. Technol.* **2014**, *48*, 13668–13674. [CrossRef]
24. Khan, S.A. Artificial sweeteners: Safe or unsafe? *J. Pak. Med. Assoc.* **2015**, *25*, 225–227.
25. García-Almeida, J.M.; Conejo-Pareja, I.M.; Muñoz-Garach, A.; Gómez-Pérez, A.; García-Alemán, J. Sweeteners: Regulatory Aspects. In *Sweeteners, Reference Series in Phytochemistry*; Mérillon, J.-M., Ramawat, K.G., Eds.; Springer International Publishing: Basel, Switzerland, 2016; ISBN 978-3-319-27026-5.
26. Shah, R.; de Jager, L.S. Recent analytical methods for the analysis of sweeteners in food: A regulatory perspective. *Food Drug Adm. Papers* **2017**, *5*, 13–31.
27. Sharma, A.; Amarnath, S.; Thulasimani, M.; Ramaswamy, S. Artificial sweeteners as a sugar substitute: Are they really safe? *Indian J. Pharmacol.* **2016**, *48*, 237–240. [CrossRef] [PubMed]
28. Vaněk, T.; Nepovím, A.; Valíček, P. Determination of stevioside in plant material and fruit teas. *J. Food Compos. Anal.* **2001**, *14*, 383–388. [CrossRef]
29. Chaturvedula, V.S.P.; Zamora, J. Reversed-Phase HPLC analysis of steviol glycosides isolated from *Stevia rebaudiana* Bertoni. *Food Nutr. Sci.* **2014**, *5*, 1711–1716. [CrossRef]
30. Wang, T.-H.; Avula, B.; Tang, W.; Wang, M.; Elsohly, M.A.; Khan, I.A. Ultra-HPLC method for quality and adulterant assessment of steviol glycosides sweeteners—*Stevia rebaudiana* and stevia products. *Food Addit. Contam. Part A* **2015**, *32*, 674–685.
31. Martono, Y.; Riyanto, S.; Rohman, A.; Martono, S. Improvement method of fast and isocratic RP-HPLC analysis of major diterpene glycoside from *Stevia rebaudiana* leaves. *AIP Conf. Proc.* **2016**, *1755*, 080001–080008.
32. Sádecka, J.; Polonský, J. Determination of inorganic ions in food and beverages by capillary electrophoresis. *J. Chromatogr.* **1999**, *834*, 401–417. [CrossRef]
33. Rios, R.V.; Pessanha, M.D.F.; de Almeida, P.F.; Viana, C.L.; Lannes, S.C. Application of fats in some food products. *Food Sci. Technol.* **2014**, *34*, 3–15. [CrossRef]
34. Li, H.; Zhou, X.; Gao, P.; Li, Q.; Li, H.; Huang, R.; Wu, M. Inhibition of lipid oxidation in foods and feeds and hydroxyl radical treated fish erythrocytes: A comparative study of *Ginkgo biloba* leaves extracts and synthetic antioxidants. *Anim. Nutr.* **2016**, *2*, 234–241. [CrossRef]
35. Kerr, B.J.; Kellner, T.A.; Shurson, G.C. Characteristics of lipids and their feeding value in swine diets. *J. Anim. Sci. Biotechnol.* **2015**, *6*, 30. [CrossRef]
36. Shurson, G.C.; Kerr, B.J.; Hanson, A.R. Evaluating the quality of feed fats and oils and their effects on pig growth performance. *J. Anim. Sci. Biotechnol.* **2015**, *6*, 10. [CrossRef]
37. Pickova, J. Importance of knowledge on lipid composition of foods to support development towards consumption of higher levels of n-3 fatty acids via freshwater fish. *Psychol. Res.* **2009**, *58*, S39–S45.
38. Benkerroum, N. Biogenic amines in dairy products: Origin, incidence, and control means. *Compr. Rev. Food Sci. Food Saf.* **2016**, *15*, 801–826. [CrossRef]
39. Spano, G.; Russo, P.; Lonvaud-Funel, A.; Lucas, P.; Alexandre, H.; Grandvalet, C.; Coton, E.; Coton, M.; Barnavon, L.; Bach, B.; et al. Biogenic amines in fermented foods. *Eur. J. Clin. Nutr.* **2010**, *64*, S95–S100. [CrossRef] [PubMed]
40. Landete, J.M.; Ferrer, S.; Polo, L.; Pardo, I. Biogenic amines in wines from three Spanish regions. *J. Agric. Food Chem.* **2005**, *53*, 1119–1124. [CrossRef]
41. Biji, K.B.; Ravishankar, C.N.; Venkateswarlu, R.; Mohna, C.O.; Srinivasa, T.K. Biogenic amines in seafood: A review. *J. Food Sci. Technol.* **2016**, *53*, 2210–2218. [CrossRef] [PubMed]

42. Karau, A.; Grayson, I. Amino acids in human and animal nutrition. *Adv. Biochem. Eng. Biotechnol.* **2014**, *143*, 189–228. [PubMed]
43. Hardy, K.; Brand-Miller, J.; Brown, K.D.; Thomas, M.G.; Copeland, L. The importance of dietary carbohydrate in human evolution. *Q. Rev. Biol.* **2015**, *90*, 251–268. [CrossRef] [PubMed]
44. Bach Knudsen, K.E.; Lærke, H.N.; Ingerslev, A.K.; Hedemann, M.S.; Nielsen, T.S.; Theil, P.K. Carbohydrates in pig nutrition—Recent advances. *J. Anim. Sci.* **2016**, *94*, 1–11. [CrossRef]
45. McDowell, L.R. *Vitamins in Animal and Human Nutrition*; Iowa State University Press: Ames, IA, USA, 2000; pp. 15–217. ISBN 0-8138-2630-6.
46. Koleva, I.I.; van Beek, T.A.; Soffers, A.E.M.F.; Dusemund, B.; Rietjens, I.M.C.M. Alkaloids in the human food chain—Natural occurrence and possible adverse effects. *Mol. Nutr. Food Res.* **2012**, *56*, 30–52. [CrossRef]
47. Di Lorenzo, C.; Dos Santos, A.; Colombo, F.; Moro, E.; Dell'Agli, M.; Restani, P. Development and validation of HPLC method to measure active amines in plant food supplements containing *Citrus aurantium* L. *Food Control* **2014**, *46*, 136–142. [CrossRef]
48. Endlová, L.; Laryšová, A.; Vrbovsky, V.; Navrátilová, Z. Analysis of alkaloids in poppy straw by high-performance liquid chromatography. *IOSR J. Eng.* **2015**, *3*, 1–7.
49. Arai, K.; Terashima, H.; Aizaewa, S.; Taga, A.; Yamamoto, A.; Tsutsumiuchi, K.; Kodama, S. Simultaneous Determination of Trigonelline, Caffeine, Chlorogenic Acid and Their Related Compounds in Instant Coffee Samples by HPLC Using an Acidic Mobile Phase Containing Octanesulfonate. *Anal. Sci.* **2015**, *31*, 831–835. [CrossRef] [PubMed]
50. Wang, H.-B.; Qi, W.; Zhang, L.; Yuan, D. Qualitative and quantitative analyses of alkaloids in *Uncaria* species by UPLC-ESI-Q-TOF/MS. *Chem. Pharm. Bull.* **2014**, *62*, 1100–1109. [CrossRef] [PubMed]
51. Ramdani, D.; Chaudhry, A.S.; Seal, C.J. Alkaloid and polyphenol analysis by HPLC in green and black tea powders and their potential use as additives in ruminant diets. *AIP Conf. Proc.* **2018**, *1927*, 0300081–0300086.
52. Yashin, A.; Yashin, Y.; Xia, X.; Nemzer, B. Chromatographic Methods for Coffee Analysis: A Review. *J. Food Res.* **2017**, *6*, 60–82. [CrossRef]
53. Pereira, C.A.M.; Rodrigues, T.R.; Yariwake, J.H. Quantification of Harman alkaloids in sour passion fruit pulp and seeds by a novel dual SBSE-LC/Flu (stir bar sorptive extraction-liquid chromatography with fluorescence detector) method. *J. Braz. Chem. Soc.* **2014**, *25*, 1472–1483. [CrossRef]
54. Kowalczyk, E.; Patyra, E.; Grelik, A.; Kwiatek, K. Development and validation of an analytical method for determination of ergot alkaloids in animal feedingstuffs with high performance liquid chromatography-fluorescence detection. *Pol. J. Vet. Sci.* **2016**, *19*, 559–565. [CrossRef]
55. Upadhyay, V.; Sharma, N.; Joshi, H.M.; Malik, A.; Mishra, M.; Singh, B.P.; Tripathi, S. Development and validation of rapid RP-HPLC method for estimation of piperine in *Piper nigrum* L. *Int. J. Herb. Med.* **2013**, *1*, 6–9.
56. Granados-Chinchilla, F.; Rodriguez, C. Tetracyclines in food and feedingstuffs: From regulation to analytical methods, bacterial resistance, and environmental and health implications. *J. Anal. Methods Chem.* **2017**, *2017*, 1315497. [CrossRef]
57. Alshannaq, A.; Yu, J.-H. Occurrence, toxicity, and analysis of major mycotoxins in food. *Int. J. Environ. Res. Public Health* **2017**, *14*, 632. [CrossRef]
58. Granados-Chinchilla, F.; Molina, A.; Chavarría, G.; Alfaro-Cascante, M.; Bogantes-Ledezma, D.; Murillo-Williams, A. Aflatoxins occurrence through the food chain in Costa Rica: Applying the One Health approach to mycotoxin surveillance. *Food Control* **2017**, *82*, 217–226. [CrossRef]
59. Oliveira, F.A.; Pereira, E.N.C.; Gobbi, J.M.; Soto-Blanco, B.; Melo, M.M. Multiresidue method for detection of pesticides in beef meat using liquid chromatography coupled to mass spectrometry detection (LC-MS) after QuEChERS extraction. *Food Addit. Contam. Part A* **2018**, *35*, 94–109. [CrossRef] [PubMed]
60. Schnabel, K.; Schmitz, R.; von Soosten, D.; Frahm, J.; Kersten, S.; Meyer, U.; Breves, G.; Hackenberg, R.; Spitzke, M.; Dänicke, S. Effects of glyphosate residues and different concentrate feed proportions on performance, energy metabolism and health characteristics in lactating dairy cows. *Arch. Anim. Nutr.* **2017**, *71*, 413–427. [CrossRef] [PubMed]
61. Velkoska-Markovska, L.; Petanovska-Ilievska, B.; Markovski, A. Application of high performance liquid chromatography to the analysis of pesticide residues in apple juice. *Contemp. Agric.* **2018**, *67*, 93–102. [CrossRef]
62. Recjczak, T.; Tuzimski, T. Application of high-performance liquid chromatography with diode array detector for simultaneous determination of 11 synthetic dyes in selected beverages and foodstuffs. *Food Anal. Methods* **2017**, *10*, 3572–3588. [CrossRef]

63. Shehata, A.B.; Rizk, M.S.; Rend, E.A. Certification of caffeine reference material purity by ultraviolet/visible spectrophotometry and high-performance liquid chromatography with diode-array detection as two independent analytical methods. *J. Food Drug Anal.* **2016**, *24*, 703–715. [CrossRef] [PubMed]
64. Cancho Grande, B.; Falcón, M.S.G.; Comesaña, M.R.; Gándara, J.S. Determination of sulfamethazine and trimethoprim in liquid feed premixes by HPLC and diode array detection, with an analysis of the uncertainty of the analytical results. *J. Agric. Food Chem.* **2001**, *49*, 3145–3150. [CrossRef]
65. Barbosa, J.; Moura, S.; Barbosa, R.; Ramos, F.; Silveira, M.I. Determination of nitrofurans in animal feeds by liquid chromatography-UV photodiode array detection and liquid chromatography-ionspray tandem mass spectrometry. *Anal. Chim. Acta* **2007**, *586*, 359–365. [CrossRef]
66. Horigome, J.; Kozuma, M.; Shirasaki, T. Fluorescence pattern analysis to assist food safety—Food analysis technology driven by fluorescence fingerprints. *Hitachi Rev.* **2016**, *65*, 248–253.
67. Borràs, S.; Companyó, R.; Guiteras, J. Analysis of sulfonamides in animal feeds by liquid chromatography with fluorescence detection. *J. Agric. Food Chem.* **2011**, *59*, 5240–5247. [CrossRef]
68. Patyra, E.; Kwiatek, K. Determination of fluoroquinolones in animal feed by ion pair high-performance liquid chromatography with fluorescence detection. *Anal. Lett.* **2017**, *50*, 1711–1720. [CrossRef]
69. Wacoo, A.P.; Wendiro, D.; Vuzi, P.C.; Hawumba, J.F. Methods for Detection of Aflatoxins in Agricultural Food Crops. *J. Appl. Chem.* **2014**, *2014*, 70629. [CrossRef]
70. Shuib, N.S.; Makahleh, A.; Salhimi, S.M.; Saad, B. Determination of aflatoxin M_1 in milk and dairy products using highperformance liquid chromatography-fluorescence with post column photochemical derivatization. *J. Chromatogr. A* **2017**, *1510*, 51–56. [CrossRef] [PubMed]
71. González de la Huebra, M.J.; Vincent, U.; von Holst, C. Sample preparation strategy for the simultaneous determination of macrolide antibiotics in animal feedingstuffs by liquid chromatography with electrochemical detection (HPLC-ECD). *J. Pharm. Biomed. Anal.* **2007**, *43*, 1628–1637. [CrossRef] [PubMed]
72. Gazdik, Z.; Ztka, O.; Petrlova, J.; Adam, V.; Zehnalek, J.; Horna, A.; Reznicek, V.; Beklova, M.; Kizek, R. Determination of vitamin C (ascorbic acid) using high performance liquid chromatography coupled with electrochemical detection. *Sensors* **2008**, *8*, 7097–7112. [CrossRef]
73. Başaran, U.; Akkbik, M.; Mut, H.; Gülümser, E.; Doğrusöz, M.C.; Koçoğlu, S. High-Performance liquid chromatography with refractive index detection for the determination of inulin in chicory roots. *Anal. Lett.* **2018**, *51*, 83–95. [CrossRef]
74. Kupina, S.; Roman, M. Determination of total carbohydrates in wine and wine-like beverages by HPLC with a refractive index detector: First action 2013.12. *J. AOAC Int.* **2014**, *97*, 498–505. [CrossRef]
75. Zhang, J.; Zhu, Y. Determination of betaine, choline and trimethylamine in feed additive by ion-exchange liquid chromatography/non-suppressed conductivity detection. *J. Chromatogr. A* **2007**, *1170*, 114–117. [CrossRef]
76. Wei, D.; Xang, X.; Wang, N.; Zhu, Y. A rapid ion chromatography column-switching method for online sample pretreatment and determination of L-carnitine, choline and mineral ions in milk and powdered infant formula. *RSC Adv.* **2017**, *7*, 5920–5927. [CrossRef]
77. Troise, A.D.; Fiore, A.; Fogliano, V. Quantitation of acrylamide in foods by high-resolution mass spectrometry. *J. Agric. Food Chem.* **2014**, *62*, 74–79. [CrossRef]
78. Patyra, E.; Nebot, C.; Gavilán, R.E.; Cepeda, A.; Kwiatek, K. Development and validation of an LC-MS/MS method for the quantification of tiamulin, trimethoprim, tylosin, sulfadiazine and sulfamethazine in medicated feed. *Food Addit. Contam. Part A* **2018**, *35*, 882–891. [CrossRef] [PubMed]
79. Wang, K.; Lin, K.; Huang, X.; Chen, M. A simple and fast extraction method for the determination of multiclass antibiotics in eggs using LC-MS/MS. Journal of Agricultural and Food Chemistry. *J. Agric. Food Chem.* **2017**, *65*, 5064–5073. [CrossRef] [PubMed]
80. Winkler, J.; Kersten, S.; Valenta, H.; Meyer, U.; Engelhardt, U.H.; Dänicke, S. Development of a multi-toxin method for investigating the carryover of zearalenone, deoxynivalenol and their metabolites into milk of dairy cows. *Food Addit. Contam. Part A* **2015**, *32*, 371–380.
81. Botha, C.J.; Legg, M.J.; Truter, M.; Sulyok, M. Multitoxin analysis of *Aspergillus clavatus*-infected feed samples implicated in two outbreaks of neuromycotoxicosis in cattle in South Africa. *Onderstepoort J. Vet. Res.* **2014**, *81*, e1–e6. [CrossRef] [PubMed]

82. Granados-Chinchilla, F.; Artavia, G. A straightforward LC approach using an amine column and single quad mass detector to determine choline chloride in feed additives and feeds. *MethodsX* **2017**, *4*, 297–304. [CrossRef] [PubMed]
83. Nassar, A.-E.F.; Bjorge, S.M. On-Line liquid chromatography-accurate radioisotope counting coupled with a radioactivity detector and mass spectrometer for metabolite identification in drug discovery and development. *Anal. Chem.* **2003**, *75*, 785–790. [CrossRef] [PubMed]
84. Kim, H.J.; Bae, I.K.; Jeong, M.H.; Park, H.J.; Jung, J.S.; Kim, J.E. A new HPLC-ELSD method for simultaneous determination of N-acetylglucosamine and N-acetylgalactosamine in dairy foods. *Int. J. Anal. Chem.* **2015**, *2015*, 892486. [CrossRef]
85. Yan, W.; Wang, N.; Zhang, P.; Zhang, J.; Wu, S.; Zhu, Y. Simultaneous determination of sucralose and related compounds by high-performance liquid chromatography with evaporative light scattering detection. *Food Chem.* **2016**, *204*, 358–364. [CrossRef]
86. Wang, Y.; Wang, M.J.; Li, J.; Yao, S.C.; Xue, J.; Zou, W.B.; Hu, C.Q. determination of spectinomycin and related substances by HPLC coupled with evaporative light scattering detection. *Acta Chromatogr.* **2015**, *207*, 93–109. [CrossRef]
87. Ligor, M.; Studzińska, S.; Horna, A.; Buszewski, B. Corona-charged aerosol detection: An analytical approach. *Crit. Rev. Anal. Chem.* **2013**, *43*, 64–78. [CrossRef]
88. Vehovec, T.; Obreza, A. Review of operating principle and applications of the charged aerosol detector. *J. Chromatogr. A* **2010**, *1217*, 1549–1556. [CrossRef]
89. Grembecka, M.; Lebiedzińska, A.; Szefer, P. Simultaneous separation and determination of erythritol, xylitol, sorbitol, mannitol, maltitol, fructose, glucose, sucrose and maltose in food products by high performance liquid chromatography coupled to charged aerosol detector. *Microchem. J.* **2014**, *117*, 77–82. [CrossRef]
90. Szekeres, A.; Budai, A.; Bencsik, O.; Németh, L.; Bartók, T.; Szécsi, Á.; Mesterházy, Á.; Vágvölgyi, C. Fumonisin measurement from maize samples by high-performance liquid chromatography coupled with corona charged aerosol detector. *J. Chromatogr. Sci.* **2014**, *52*, 1181–1185. [CrossRef]
91. Enri, F.; Steuer, W.; Bosshart, H. Automation and validation of HPLC-Systems. *Chromatographia* **1987**, *24*, 201–207.
92. Oláh, E.; Tarnai, M.; Fekete, J. Possibility of large volume injection and band focusing in UHPLC. *J. Chromatogr. Sci.* **2013**, *51*, 839–844. [CrossRef] [PubMed]
93. Williamson, G. The role of polyphenols in modern nutrition. *Nutr. Bull.* **2017**, *42*, 226–235. [CrossRef]
94. Pandey, K.B.; Rizvi, S.I. Plant polyphenols as dietary antioxidants in human health and disease. *Oxidaive Med. Cell. Longev.* **2009**, *2*, 270–278. [CrossRef] [PubMed]
95. Beaulieu, M.; Franke, K.; Fischer, K. Feeding on ripening and over-ripening fruit: Interactions between sugar, ethanol and polyphenol contents in a tropical butterfly. *J. Exp. Biol.* **2017**, *220*, 3127–3134. [CrossRef]
96. Vanholme, R.; Demedts, B.; Morreel, K.; Ralph, J.; Boerjan, W. Lignin Biosynthesis and Structure. *Plant Physiol.* **2010**, *153*, 895–905. [CrossRef]
97. Bensalem, J.; Dal-Pen, A.; Gillard, E.; Fréderic, C.; Pallet, V. Protective effects of berry polyphenols against age-related cognitive impairment. *Nutr. Aging* **2015**, *3*, 89–106. [CrossRef]
98. Kowalska, K.; Olejnik, A.; Szwajgier, D.; Olkowicz, M. Inhibitory activity of chokeberry, bilberry, raspberry and cranberry polyphenol-rich extract towards adipogenesis and oxidative stress in differentiated 3T3-L1 adipose cells. *PLoS ONE* **2017**, *12*, e0188583. [CrossRef] [PubMed]
99. Khalifa, I.; Zhu, W.; Li, K.-K.; Li, C.-M. Polyphenols of mulberry fruits as multifaceted compounds: Compositions, metabolism, health benefits, and stability—A structural review. *J. Funct. Foods* **2018**, *40*, 28–43. [CrossRef]
100. Pérez-Jiménez, J.; Neveu, V.; Vos, F.; Sclbert, A. Identification of the 100 richest dietary sources of polyphenols: An application of the Phenol-Explorer database. *Eur. J. Clin. Nutr.* **2010**, *64*, S112–S120. [CrossRef] [PubMed]
101. Acosta, O.; Vaillant, F.; Pérez, A.M.; Dornier, M. Concentration of polyphenolic compounds in blackberry (*Rubus adenotrichos* Schltdl) juice by nanofiltration. *J. Food Process Eng.* **2017**, *40*, e12343. [CrossRef]
102. Haminiuk, C.W.; Maciel, G.M.; Plata-Oviedo, M.S.V.; Peralta, R.M. Phenolic compounds in fruits—An overview. *Int. J. Food Sci. Technol.* **2012**, *47*, 2023–2044. [CrossRef]
103. Karasawa, M.M.G.; Mohan, C. Fruits as Prospective Reserves of bioactive Compounds: A Review. *Nat. Prod. Bioprospect.* **2018**, *8*, 335–346. [CrossRef]
104. Reynoso-Camacho, R.; Rufino, M.S.M.; Amaya-Cruz, D.M.; Pérez, A.M. Non-extractable polyphenols in tropical fruits: Occurrence and health-related properties. In *Non-extractable Polyphenols and Carotenoids:*

Importance in Human Nutrition and Health; Saura-Calixto, F., Pérez-Jiménez, J., Eds.; Royal Society of Chemistry: Cambridge, UK, 2018; ISBN 978-1-78801-106-8.

105. Abbas, M.; Saeed, F.; Anjum, F.M.; Afzaal, M.; Tufail, T.; Bashir, M.S.; Ishtiaq, A.; Hussain, S.; Suleria, H.A. Natural polyphenols: An overview. *Int. J. Food Prop.* **2016**, *20*, 1689–1699. [CrossRef]
106. Kaşikci, M.B.; Bağdatlioğlu, N. High hydrostatic pressure treatment of fruit, fruit products and fruit juices: A review on phenolic compounds. *J. Food Health Sci.* **2016**, *2*, 27–39.
107. Miletić, N.; Mitrović, O.; Popović, B.; Nedović, V.; Zlatković, B.; Kandić, M. Polyphenolic content and antioxidant capacity in fruits of plum (*Prunus domestica* L.) cultivars "Valjevka" and "Mildora" as influenced by air drying. *J. Food Qual.* **2013**, *36*, 229–237. [CrossRef]
108. McSweeny, M.; Seetharaman, K. State of polyphenols in the drying process of fruits and vegetables. *Crit. Rev. Food Sci. Nutr.* **2015**, *55*, 660–669. [CrossRef]
109. Koffi, E.; Sea, T.; Dodehe, Y.; Soro, S. Effect of solvent type on extraction of polyphenols from twenty three Ivorian plants. *J. Anim. Plant Sci.* **2010**, *5*, 550–558.
110. Flores, G.; Dastmalchi, K.; Wu, S.-B.; Whalen, K.; Dabo, A.J.; Reynertson, K.A.; Foronjy, R.F.; D'Armiento, J.M.; Kennelly, E.J. Phenolic-rich extract from the Costa Rican Guava (*Psidium friedrichsthalianum*) pulp with antioxidant and anti-inflammatory activity. Potential for COPD therapy. *Food Chem.* **2013**, *141*, 889–895. [CrossRef] [PubMed]
111. Tresserra-Rimbau, E.; Arranz, S.; Vallverdu-Queralt, A. New Insights into the Benefits of Polyphenols in Chronic Diseases. *Oxidative Med. Cell. Longev.* **2017**, *2017*, 1432071. [CrossRef] [PubMed]
112. Yang, J.; Dwyer, J. Exploring Possible Health Effects of Polyphenols in Foods. *Nutr. Today* **2017**, *52*, 62–72. [CrossRef]
113. Gordon, A.; Jungfer, E.; da Silva, B.A.; Maia, J.G.S.; Marx, F. Phenolic constituents and antioxidant capacity of four underutilized fruits from the Amazon region. *J. Agric. Food Chem.* **2011**, *59*, 7688–7699. [CrossRef] [PubMed]
114. Assefa, A.D.; Jeong, Y.-J.; Kim, D.-J.; Jeon, Y.-A.; Ok, H.-C.; Baek, H.-J.; Sung, J.-S. Characterization, identification, and quantification of phenolic compounds using UPLC-Q-TOF-MS and evaluation of antioxidant activity of 73 *Perilla frutescens* accessions. *Food Res. Int.* **2018**, *111*, 153–167. [CrossRef]
115. Anton, D.; Bender, I.; Kaart, T.; Roasto, M.; Heinonen, M. Changes in polyphenols contents and antioxidant capacities of organically and conventionally cultivated tomato (*Solanum lycopersicum* L.) fruits during ripening. *Int. J. Anal. Chem.* **2017**, *2017*, 2367453. [CrossRef]
116. Radovanović, B.C.; Anđelković, A.S.M.; Radovanović, A.B.; Anđelković, M.Z. Antioxidant and antimicrobial activity of polyphenol extracts from wild berry fruits grown in southeast Serbia. *Trop. J. Pharm. Res.* **2013**, *12*, 813–819. [CrossRef]
117. Veljković, J.N.; Pavlović, A.N.; Mitić, S.S.; Tošić, S.B.; Stojanović, G.S.; Kaličanin, B.M.; Stanović, D.M.; Stojković, M.B.; Mitić, M.N.; Brcanović, J.M. Evaluation of individual phenolic compounds and antioxidant properties of black, green, herbal and fruit tea infusions consumed in Serbia: Spectrophotometrical and electrochemical approaches. *J. Food Nutr. Res.* **2013**, *52*, 12–24.
118. Miletić, N.; Popović, B.; Mitrović, O.; Kandić, M.; Leposavić, A. Phenolic compounds and antioxidant capacity of dried and candied fruits commonly consumed in Serbia. *Czech J. Food Sci.* **2014**, *32*, 360–368. [CrossRef]
119. Rojas-Garbanzo, C.; Zimmermann, B.F.; Schulze-Kaysers, N.; Schieber, A. Characterization of phenolic and other polar compounds in peel and flesh of pink guava (*Psidium guajava* L. cv. 'Criolla') by ultra-high performance liquid chromatography with diode array and mass spectrometric detection. *Food Res. Int.* **2017**, *100*, 445–453. [CrossRef] [PubMed]
120. Azofeifa, G.; Quesada, S.; Pérez, A.M.; Vaillant, F.; Michel, A. Effect of an in vitro digestion on the antioxidant capacity of a microfiltrated blackberry juice (*Rubus adenotrichos*). *Beverages* **2018**, *4*, 30. [CrossRef]
121. Georgé, S.; Brat, P.; Alter, P.; Amiot, M.J. Rapid determination of polyphenols and vitamin C in plant derived products. *J. Agric. Food Chem.* **2005**, *53*, 1370–1373. [CrossRef] [PubMed]
122. Gouvêa, A.; de Araujo, M.C.; Schulz, D.F.; Pacheco, S.; Godoy, R.; Cabral, L. Anthocyanins standards (cyanidin-3-*O*-glucoside and cyanidin-3-*O*-rutinoside) isolation from freeze-dried açaí (*Euterpe oleraceae* Mart.) by HPLC. *Ciênc. Tecnol. Aliment.* **2012**, *32*, 43–46. [CrossRef]
123. Teboukeu, G.B.; Djikeng, F.T.; Klang, M.J.; Ndomou, S.H.; Karuna, M.S.L.; Womeni, H.M. Polyphenol antioxidants from cocoa pods: Extraction optimization, effect of the optimized extract, and

124. Ryu, W.-K.; Kim, H.-W.; Kim, G.-D.; Rhee, H.-I. Rapid determination of capsaicinoids by colorimetric method. *J. Food Drug Anal.* **2017**, *25*, 798–803. [CrossRef]
125. Collins, M.D.; Wasmund, L.M.; Bosland, P.W. Improved method for quantifying capsaicinoids in *Capsicum* using high-performance liquid chromatography. *Hortic. Sci.* **1995**, *30*, 137–139.
126. Guzmán, I.; Bosland, P.W. Sensory properties of chile pepper heat e and its importance to food quality and cultural preference. *Appetite* **2017**, *117*, 186–190. [CrossRef]
127. Kobata, K.; Sugawara, M.; Mimura, M.; Yawaza, S.; Watanabe, T. Potent Production of Capsaicinoids and Capsinoids by Capsicum Peppers. *J. Agric. Food Chem.* **2013**, *61*, 11127–11132. [CrossRef]
128. Garcés-Claver, A.; Gil-Ortega, R.; Álvarez-Fernández, A.; Arnedo-Andrés, M.S. Inheritance of capsaicin and dihydrocapsaicin, determined by HPLC-ESI/MS, in an intraspecific cross of *Capsicum annuum* L. *J. Agric. Food Chem.* **2007**, *55*, 6951–6957. [CrossRef]
129. Goll, J.; Frey, A.; Minceva, M. Study of the separation limits of continuous solid support free liquid-liquid chromatography: Separation of capsaicin and dihydrocapsaicin by centrifugal partition chromatography. *J. Chromatogr. A* **2013**, *1284*, 59–68. [CrossRef] [PubMed]
130. Othman, Z.A.A.; Ahmed, Y.B.H.; Habila, M.A.; Ghafar, A.A. Determination of capsaicin and dihydrocapsaicin in *Capsicum* fruit samples using high performance liquid chromatography. *Molecules* **2011**, *16*, 8919–8929. [CrossRef] [PubMed]
131. Ma, F.; Yang, Q.; Matthäus, B.; Li, P.; Zhang, Q.; Zhang, L. Simultaneous determination of capsaicin and dihydrocapsaicin for vegetable oil adulteration by immunoaffinity chromatography cleanup coupled with LC-MS/MS. *J. Chromatogr. B* **2016**, *1021*, 137–144. [CrossRef] [PubMed]
132. Zhou, C.; Ma, D.; Cao, W.; Shi, H.; Jiang, Y. Fast simultaneous determination of capsaicin, dihydrocapsaicin and nonivamide for detecting adulteration in edible and crude vegetable oils by UPLC-MS/MS. *Food Addit. Contam. Part A* **2018**, *35*, 1447–1452. [CrossRef] [PubMed]
133. Schmidt, A.; Fiechter, G.; Fritz, E.-M.; Mayer, H.K. Quantitation of capsaicinoids in different chilies from Austria by a novel UPLC method. *J. Food Compos. Anal.* **2017**, *60*, 32–37. [CrossRef]
134. Sganzerla, M.; Coutinho, J.P.; Tavares de Melo, A.M.; Godoy, E.T. Fast method of capsaicinoids analysis from *Capsicum chinense* fruits. *Food Res. Int.* **2014**, *64*, 718–725. [CrossRef]
135. Dang, Y.M.; Hong, Y.S.; Lee, C.L.; Khan, N.; Park, S.; Jeong, S.-W.; Kim, K.S. Determination of Capsaicinoids in Red Pepper Products from South Korea by High-Performance Liquid Chromatography with Fluorescence Detection. *Anal. Lett.* **2018**, *51*, 1291–1303. [CrossRef]
136. Lu, M.; Ho, C.-T.; Huang, Q. Extraction, bioavailability, and bioefficacy of capsaicinoids. *J. Food Drug Anal.* **2017**, *25*, 27–36. [CrossRef] [PubMed]
137. Zhang, Q.; Hu, J.; Sheng, L.; Li, Y. Simultaneous quantification of capsaicin and dihydrocapsaicin in rat plasma using HPLC coupled with tandem mass spectrometry. *J. Chromatogr. B* **2010**, *878*, 2292–2297. [CrossRef]
138. Kuzma, M.; Fodor, K.; Maász, G.; Avar, P.; Mózsik, G.; Past, T.; Fischer, E.; Perjési, P. A validated HPLC-FLD method for analysis of intestinal absorption and metabolism of capsaicin and dihydrocapsaicin in the rat. *J. Pharm. Biomed. Anal.* **2015**, *103*, 59–66. [CrossRef]
139. Baskaran, P.; Krishnan, V.; Ren, J.; Thyagarjan, B. Capsaicin induces browning of white adipose tissue and counters obesity by activating TRPV1 channel-dependent mechanisms. *Br. J. Pharmacol.* **2016**, *173*, 2369–2389. [CrossRef]
140. Wolde, T. Effects of caffeine on health and nutrition: A Review. *Food Sci. Qual. Manag.* **2014**, *30*, 59–65.
141. Martínez-Pinilla, E.; Oñatibia-Asibia, A.; Franco, R. The relevance of theobromine for the beneficial effects of cocoa consumption. *Front. Pharmacol.* **2015**, *6*, 1–5. [CrossRef]
142. Ahluwalia, N.; Herrick, K. Caffeine intake from food and beverage sources and trends among children and adolescents in the United States: Review of national quantitative studies from 1999 to 2011. *Adv. Nutr.* **2015**, *6*, 102–111. [CrossRef] [PubMed]
143. Verster, J.C.; Koenig, J. Caffeine intake and its sources: A review of national representative studies. *Crit. Rev. Food Sci. Nutr.* **2018**, *58*, 1250–1259. [CrossRef] [PubMed]

144. Li, X.; Simmons, R.; Mwongela, S.M. Two-stage course-embedded determination of caffeine and related compounds by HPLC in caffeine containing food, beverages and (or) related products. *J. Lab. Chem. Educ.* **2017**, *5*, 19–25.
145. Lowry, J.A.; Pearce, R.E.; Gaedigk, A.; Venneman, M.; Talib, N.; Shaw, P.; Leeder, J.S.; Kearns, G.L. Lead and its effects on cytochromes P450. *J. Drug Metab. Toxicol.* **2012**, *S5*, S004. [CrossRef]
146. Grujić-Letić, N.; Rakić, B.; Šefer, E.; Milanović, M.; Nikšić, M.; Vujić, I.; Milić, N. Quantitative determination of caffeine in different matrices. *Maced. Pharm. Bull.* **2016**, *62*, 77–84.
147. Gonçalves, E.S.; Rodrigues, S.V.; Vieira da Silva-Filho, E. The use of caffeine as a chemical marker of domestic wastewater contamination in surface waters: Seasonal and spatial variations in Teresópolis, Brazil. *Rev. Ambient. Agua* **2017**, *12*, 192–202. [CrossRef]
148. Gliszczyńska-Świgło, A.; Rybicka, I. Simultaneous determination of caffeine and water-soluble vitamins in energy drinks by HPLC with photodiode array and fluorescence detection. *Food Anal. Methods* **2015**, *8*, 139–146. [CrossRef]
149. Mazdeh, F.Z.; Moradi, Z.; Moghaddam, G.; Moradi-Khatoonabadi, Z.; Aftabdari, F.E.; Badaei, P.; Hajimahmoodi, M. Determination of synthetic food colors, caffeine, sodium benzoate and potassium sorbate in sports drinks. *Trop. J. Pharm. Res.* **2016**, *15*, 183–188. [CrossRef]
150. Ortega, N.; Romero, M.-P.; Macià, A.; Reguant, J.; Anglès, N.; Morelló, J.-R.; Motilva, M.-J. Comparative study of UPLC-MS/MS and HPLC-MS/MS to determine procyanidins and alkaloids in cocoa samples. *J. Food Compos. Anal.* **2010**, *23*, 298–305. [CrossRef]
151. Rodríguez-Carrasco, Y.; Gaspari, A.; Graziani, G.; Santini, A.; Ritieni, A. Fast analysis of polyphenols and alkaloids in cocoa-based products by ultra-high performance liquid chromatography and orbitral high resolution mass spectrometry (UHPLC-Q-Orbitrap-MS/MS). *Food Res. Int.* **2018**, *111*, 229–236. [CrossRef] [PubMed]
152. Oellig, C.; Schunck, J.; Schwack, W. Determination of caffeine, theobromine and theophylline in Mate beer and Mate soft drinks by high-performance thin-layer chromatography. *J. Chromatogr. A* **2018**, *1533*, 208–212. [CrossRef]
153. Alvi, S.N.; Hammami, M.M. Validated HPLC method for determination of caffeine level in human plasma using synthetic plasma: Application to bioavailability studies. *J. Chromatogr. Sci.* **2011**, *49*, 292–296. [CrossRef]
154. Begas, E.; Kouvaras, E.; Tsakalof, A.K.; Bounitsi, M.; Asprodini, E.K. Development and validation of a reverse-phase HPLC method for CYP1A2 phenotyping by use of a caffeine metabolite ratio in saliva. *Biomed. Chromatogr.* **2015**, *29*, 1657–1663. [CrossRef]
155. Rodríguez, A.; Costa-Bauza, A.; Saenz-Torres, C.; Rodrigo, D.; Grases, F. HPLC method for urinary theobromine determination: Effect of consumption of cocoa products on theobromine urinary excretion in children. *Clin. Biochem.* **2015**, *48*, 1138–1143. [CrossRef]
156. Shrestha, S.; Rijal, S.K.; Pokhrel, P.; Rai, K.P. A simple HPLC method for determination of caffeine content in tea and coffee. *J. Food Sci. Technol.* **2016**, *9*, 74–78.
157. Fajara, B.E.P.; Susanti, H. HPLC determination of caffeine in coffee beverage. *IOP Conf. Ser. Mater. Sci. Eng.* **2017**, *259*, 01211. [CrossRef]
158. Aşçi, B.; Zor, Ş.D.; Dönmez, Ö.A. Development and validation of HPLC method for the simultaneous determination of five food additives and caffeine in soft drinks. *Int. J. Anal. Chem.* **2016**, *2016*, 2879406. [CrossRef]
159. Yamamoto, T.; Takahashi, H.; Suzuki, K.; Hirano, A.; Kamei, M.; Goto, T.; Takahashi, N.; Kawada, T. Theobromine enhances absorption of cacao polyphenol in rats. *Biosci. Biotechnol. Biochem.* **2014**, *78*, 2059–2063. [CrossRef] [PubMed]
160. Romano, R.; Santini, A.; Le Grottaglie, L.; Manzo, N.; Visconti, A.; Ritieni, A. Identification markers based on fatty acid composition to differentiate between roasted *Arabica* and *Canephora (Robusta)* coffee varieties in mixtures. *J. Food Compos. Anal.* **2014**, *35*, 1–9. [CrossRef]
161. López-Sánchez, R.C.; Lara-Díaz, V.J.; Aranda-Gutiérrez, A.; Martínez-Cardona, J.A.; Hernández, J.A. HPLC method for quantification of caffeine and its three major metabolites in human plasma using fetal bovine serum matrix to evaluate prenatal drug exposure. *J. Anal. Methods Chem.* **2018**, *2018*, 2085059. [CrossRef] [PubMed]
162. Ptolemy, A.S.; Tzioumis, E.; Thomke, A.; Rifai, S.; Kellogg, M. Quantification of theobromine and caffeine in saliva, plasma and urine via liquid chromatography-tandem mass spectrometry: A single analytical protocol applicable to cocoa intervention studies. *J. Chromatogr. A* **2010**, *878*, 409–416. [CrossRef] [PubMed]

163. Kobayashi, J.; Ikeda, K.; Terada, H.; Mochizuki, M.; Sugiyama, H. HPLC determination of caffeine using a photodiode array detector and applying a derivative processing to chromatograms. *Bull. Nippon Vet. Life Sci. Univ.* **2014**, *63*, 48–57.
164. Naviglio, D.; Gallo, M.; Le Grottaglie, L.; Scala, C.; Ferrara, L.; Santini, A. Determination of cholesterol in Italian chicken eggs. *Food Chem.* **2012**, *132*, 701–708. [CrossRef]
165. Ribeiro, S.M.L.; Luz, S.S.; Aquino, R.S. The role of nutrition and physical activity in cholesterol and aging. *Clin. Geriatr. Med.* **2015**, *31*, 401–416. [CrossRef]
166. Albuquerque, T.G.; Oliveira, M.B.P.P.; Sanches-Silva, A.; Costa, H.S. Cholesterol determination in foods: Comparison between high performance and ultra-high performance liquid chromatography. *Food Chem.* **2016**, *193*, 18–25. [CrossRef] [PubMed]
167. Gylling, H.; Simonen, P. Are plant sterols and plant stenols are a viable future treatment for dyslipidemia? *Expert Rev. Cardiovasc. Ther.* **2016**, *4*, 549–551. [CrossRef] [PubMed]
168. Sonawane, P.D.; Pollier, J.; Panda, S.; Szymanski, J.; Massalha, H.; Yona, M.; Unger, T.; Malitsky, S.; Arendt, P.; Powels, L.; et al. Plant cholesterol biosynthetic pathway overlaps with phytosterol metabolism. *Nat. Plants* **2016**, *3*, 16205. [CrossRef] [PubMed]
169. Cruz, R.; Casal, S.; Mendes, E.; Costa, A.; Santos, C.; Morais, S. Validation of a single-extraction procedure for sequential analysis of vitamin E, cholesterol, fatty acids, and total fat in seafood. *Food Anal. Methods* **2012**, *6*, 1196–1204. [CrossRef]
170. Saldanha, T.; Bragagnolo, N. Effect of grilling on cholesterol oxide formation and fatty acids alterations in fish. *Ciênc. Tecnol. Aliment.* **2010**, *30*, 385–390. [CrossRef]
171. Bauer, L.C.; Santana, D.A.; Macedo, M.S.; Torres, A.G.; de Souza, N.E.; Simionato, J.I. Method validation for simultaneous determination of cholesterol and cholesterol oxides in milk by RP-HPLC-DAD. *J. Braz. Chem. Soc.* **2014**, *25*, 161–168. [CrossRef]
172. Daneshfar, A.; Khezeli, T.; Lotfi, H.J. Determination of cholesterol in food samples using dispersive liquid-liquid micro extraction followed by HPLC-UV. *J. Chromatogr. B* **2009**, *877*, 456–460. [CrossRef]
173. Georgiou, C.A.; Constantinou, M.S.; Kapnissi-Christodoulou, C.P. Sample preparation: A critical step in the analysis of cholesterol oxidation products. *Food Chem.* **2014**, *145*, 1918–1926. [CrossRef]
174. Robinet, P.; Wang, Z.; Hazen, S.L.; Smith, J.D. A simple and sensitive enzymatic method for cholesterol quantification in macrophages and foam cells. *J. Lipid Res.* **2010**, *51*, 3364–3369. [CrossRef]
175. Carvalho, P.O.; Campos, P.R.B.; Noffs, M.D.; Fegolente, P.B.L.; Fegolente, L.V. Enzymatic hydrolysis of salmon oil by native lipases: Optimization of process parameters. *J. Braz. Chem. Soc.* **2009**, *20*, 117–124. [CrossRef]
176. Codex Alimentarius. Code of Practice for the Prevention and Reduction of Mycotoxin Contamination in Cereals (CAC/RCP 51-2003). 2003. Available online: http://www.fao.org/fao-who-codexalimentarius/codex-texts/codes-of-practice/es/ (accessed on 20 June 2018).
177. Njumbe Ediage, E.; Van Poucke, C.; De Saeger, S. A multi-analyte LC-MS/MS method for the analysis of 23 mycotoxins in different sorghum varieties: The forgotten sample matrix. *Food Chem.* **2015**, *117*, 397–404. [CrossRef]
178. Binder, E. Managing the risk of mycotoxins in modern feed production. *Anim. Feed Sci. Technol.* **2007**, *133*, 149–166. [CrossRef]
179. Dzuman, Z.; Zachariasova, M.; Lacina, O.; Veprokova, Z.; Slavokova, P.; Hajslova, J. A rugged high-throughput analytical approach for the determination and quantification of multiple mycotoxins in complex feed matrices. *Talanta* **2014**, *121*, 263–272. [CrossRef]
180. Chavarría, G.; Molina, A.; Leiva, A.; Méndez, G.; Wong-González, E.; Cortés-Muñoz, M.; Rodríguez, C.; Granados-Chinchilla, F. Distribution, stability, and protein interactions of Aflatoxin M1 in fresh cheese. *Food Control* **2017**, *73*, 581–586. [CrossRef]
181. Ostry, V.; Malir, F.; Toman, J.; Grosse, Y. Mycotoxins as human carcinogens, the IARC Monographs classification. *Mycotoxins Res.* **2016**, *33*, 65–73. [CrossRef]
182. CAST. *Mycotoxins: Risks in Plant, Animal, and Human Systems*; Council for Agricultural Science and Technology: Ames, IA, USA, 2003; pp. 13–48. ISBN 1-887383-22-0.
183. Molina Alvarado, A.; Zamora-Sanabria, R.; Granados-Chinchilla, F. A Focus on Aflatoxins in Feedstuffs: Levels of Contamination, Prevalence, Control Strategies, and Impacts on Animal Health. In *Aflatoxin Control, Analysis, Detection and Health Risks*; Lukman Bola Abdulra'uf, Ed.; IntechOpen: London, UK, 2017; pp. 115–152.

184. Codex Alimentarius. Code of Practice for the Reduction of Aflatoxin B_1 in Raw Materials and Supplemental Feedingstuffs for Milk Producing Animals (CAC/RCP 45-1997). 1997. Available online: http://www.fao.org/fao-who-codexalimentarius/codex-texts/codes-of-practice/es/ (accessed on 20 June 2018).
185. Flores-Flores, M.; González-Peñas, E. Development and validation of a high performance liquid chromatographic-mass spectrometry method for the simultaneous quantification of 10 trichothecenes in ultra-high temperature processed cow milk. *J. Chromatogr. A* **2015**, *1419*, 37–44. [CrossRef]
186. Njumbe Ediage, E.; Diana Di Mavungu, J.; Monbaliu, S.; Van Peteghem, C.; De Saeger, S. A Validated Multianalyte LC–MS/MS Method for Quantification of 25 Mycotoxins in Cassava Flour, Peanut Cake and Maize Samples. *J. Agric. Food Chem.* **2011**, *59*, 5173–5180. [CrossRef] [PubMed]
187. Rasmussen, R.; Storm, I.; Rasmussen, P.; Smedsgaard, J.; Nielsen, K. Multi-mycotoxin analysis of maize silage by LC-MS/MS. *Anal. Bioanal. Chem.* **2010**, *397*, 765–776. [CrossRef]
188. Chala, A.; Taye, W.; Ayalew, A.; Krska, R.; Sulyok, M.; Logrieco, A. Multimycotoxin analysis of sorghum (Sorghum bicolor L. Moench) and finger millet (*Eleusine coracana* L. Gaten) from Ethiopia. *Food Control* **2014**, *45*, 29–35. [CrossRef]
189. Vishwanath, V.; Sulyok, M.; Labuda, R.; Bicker, W.; Krska, R. Simultaneous determination of 186 fungal and bacterial metabolites in indoor matrices by liquid chromatography/tandem mass spectrometry. *Anal. Bioanal. Chem.* **2009**, *395*, 1355–1372. [CrossRef]
190. Alija, C.M.; Brar, S.K.; Verma, M.; Tyagi, R.D.; Godbout, S.; Valéro, J.R. Bio-processing of agro-byproducts to animal feed. *Crit. Rev. Biotechnol.* **2012**, *32*, 382–400.
191. Mikušová, P.; Ritieni, A.; Santini, A.; Juhasová, G.; Šrobárová, A. Contamination by mould of grape berries in Slovakia. *Food Addit. Contam. Part A* **2010**, *27*, 738–747. [CrossRef]
192. Santini, A.; Ferracane, R.; Meca, G.; Ritieni, A. Overview of analytical methods for beauvericin and fusaproliferin in food matrices. *Anal. Bioanal. Chem.* **2009**, *395*, 1253–1260. [CrossRef]
193. Gil-Serna, J.; Vázquez, C.; González-Jaén, M.T.; Patiño, B. Wine contamination with ochratoxins: A Review. *Beverages* **2018**, *4*, 6. [CrossRef]
194. Pizzutti, I.R.; de Kok, A.; Scholten, J.; Righi, L.W.; Cardoso, C.D.; Rohers, G.N.; da Silva, R.C. Development, optimization and validation of a multimethod for the determination of 36 mycotoxins in wines by liquid chromatography–tandem mass spectrometry. *Talanta* **2014**, *129*, 352–363. [CrossRef] [PubMed]
195. Nistor, E.; Dobre, A.; Dobre, A.; Bampidis, V.; Ciola, V. Grape pomace in sheep and dairy cows feeding. *J. Hortic. For. Biotechnol.* **2014**, *18*, 146–150.
196. Guerra-Rivas, C.; Gallardo, B.; Mantecón, Á.R.; del Álamo-Sanza, M.; Manso, T. Evaluation of grape pomace from red winw by-products as feed for sheep. *J. Sci. Food Agric.* **2017**, *97*, 1885–1893. [CrossRef] [PubMed]
197. Kerasioti, E.; Terzopoulou, Z.; Komini, O.; Kafantaris, I.; Makri, S.; Stagos, D.; Gerasopoulos, K.; Anisimov, N.Y.; Tsatsakis, A.M.; Kouretas, D. Tissue specific effects of feeds supplemented with grape pomace or olive oil mill wastewater on detoxification enzymes in sheep. *Toxicol. Rep.* **2017**, *4*, 364–372. [CrossRef] [PubMed]
198. Avantaggiato, G.; Greco, D.; Damascelli, A.; Solfrizzo, M.; Visconti, A. Assessment of Multi-mycotoxin Adsorption Efficacy of Grape Pomace. *J. Agric. Food Chem.* **2014**, *62*, 497–507. [CrossRef] [PubMed]
199. Gambacorta, L.; Pinton, P.; Avantaggiato, G.; Oswald, I.P.; Solfrizzo, M. Grape Pomace, an Agricultural Byproduct Reducing Mycotoxin Absorption: In Vivo Assessment in Pig Using Urinary Biomarkers. *J. Agric. Food Chem.* **2016**, *64*, 6762–6771. [CrossRef]
200. Anadón, A.; Martínez-Larrañaga, M.; Ares, I.; Martínez, M. Chapter 7—Regulatory Aspects for the Drugs and Chemicals Used in Food-Producing Animals in the European Union. In *Veterinary Toxicology: Basic and Clinical Principles*, 3rd ed.; Gupta, R.C., Ed.; Academic Press: Cambridge, MA, USA, 2018; pp. 103–131.
201. Decheng, S.; Peilong, W.; Yang, L.; Ruiguo, W.; Shulin, W.; Zhiming, X.; Su, Z. Simultaneous determination of antibiotics and amantadines in animal-derived feedstuffs by ultraperformance liquid chromatographic-tandem mass spectrometry. *J. Chromatogr. B* **2018**, *1095*, 183–190. [CrossRef]
202. Molognoni, L.; Coelho de Souza, N.; Antunes de Sá Ploêncio, L.; Amadeu Micke, G.; Daguer, H. Simultaneous analysis of spectinomycin, halquizol, zilpaterol, and melamine in feedingstuffs by ion-pair liquid chromatography-tandem mass spectrometry. *J. Chromatogr. A* **2018**, *1569*, 110–117. [CrossRef]
203. Cancho Grande, B.; García Falcón, M.S.; Simal Gándara, J. El uso de los antibióticos en la alimentación animal: Perspectiva actual. *Cienc. Tecnol. Aliment.* **2000**, *3*, 39–47.

204. Rojek-Podgórska, B. EU Legislation in Progress: Review of Medicated feed Legislation. European Parliamentary Research Service (EPRS). 2016. Available online: http://www.europarl.europa.eu/RegData/etudes/BRIE/2016/583843/EPRS_BRI%282016%29583843_EN.pdf (accessed on 10 July 2018).
205. Kang, H.; Lee, S.; Shin, D.; Jeong, J.; Hong, J.; Rhee, G. Occurrence of veterinary drug residues in farmed fishery products in South Korea. *Food Control* **2018**, *85*, 57–65. [CrossRef]
206. Kumar Saxena, S.; Rangasamy, R.; Krishnan, A.; Singh, D.; Uke, S.; Kumar Malekadi, P.; Sengar, A.; Peer Mohamed, D. Simultaneous determination of multi-residue and multi-class antibiotics in aquaculture shrimps by UPLC-MS/MS. *Food Chem.* **2018**, *260*, 336–343. [CrossRef] [PubMed]
207. Barreto, F.; Ribeiro, C.; Barcellos Hoft, R.; Dalla Costa, T. A simple and high-throughput method for determination and confirmation of 14 coccidiostats in poultry muscle eggs using liquid chromatography-quadrupole linear ion trap–tandem mass spectrometry (HPLC-QqLIT-MS/MS): Validation according to European Union 2002/657/EC. *Talanta* **2017**, *168*, 43–51. [PubMed]
208. World Health Organization. Global Action Plan on Antimicrobial Resistance. WHO Library Cataloguing 2015. Available online: http://www.wpro.who.int/entity/drug_resistance/resources/global_action_plan_eng.pdf (accessed on 18 July 2018).
209. Camargo Valese, A.; Molognoni, L.; Coelho de Souza, N.; Antunes de Sá Ploêncio, L.; Oliveira Costa, A.; Barreto, F. Development, validation and different approaches for the measurement uncertainty of a multi-class veterinary drugs residues LC-MS method for feeds. *J. Chromatogr. B* **2017**, *1053*, 48–59. [CrossRef] [PubMed]
210. European Commission. Directive 2001/82/EC of the European Parliament and of the Council of 6 November 2001 on the Community code relating to veterinary medicinal products. *Off. J. Eur. Communities* **2001**, *L 311*, 1–66. Available online: https://eur-lex.europa.eu/legal-content/EN/TXT/PDF/?uri=CELEX:32001L0082&from=ES (accessed on 10 July 2018).
211. European Commission. Regulation (EC) No 726/2004 of the European Parliament and of the Council of 31 March 2004 laying down Community procedures for the authorization and supervision of medicinal products for human and veterinary use and establishing a European Medicines Agency. *Off. J. Eur. Communities* **2004**, *L 136*, 1–33. Available online: https://eur-lex.europa.eu/legal-content/EN/TXT/PDF/?uri=CELEX:32004R0726&from=ES (accessed on 10 July 2018).
212. US Food and Drug Administration. Medicated Feeds. Available online: https://www.fda.gov/AnimalVeterinary/Products/AnimalFoodFeeds/MedicatedFeed/default.htm#license (accessed on 19 July 2018).
213. Robert, C.; Basseur, P.-Y.; Dubois, M.; Delahaut, P.; Gillard, N. Development and validation of rapid multiresidue and multi-class analysis for antibiotics and anthelmintics in feed by ultra high performance liquid chromatography coupled to tandem mass spectrometry. *Food Addit. Contam. Part A* **2016**, *33*, 1312–1323. [CrossRef]
214. Shendy, A.; Al-Ghobashy, M.; Gad Alla, S.; Lotfy, H. Development and validation of a modified QuEChERS protocol coupled to LC-MS/MS for simultaneous determination of multi-class antibiotic residues in honey. *Food Chem.* **2016**, *190*, 982–989. [CrossRef] [PubMed]
215. Wang, Y.; Xiao, C.; Guo, J.; Yuan, Y.; Wang, J.; Liu, L.; Yue, T. Development and application of method for the analysis of 9 mycotoxins in maize by HPLC-MS/MS. *J. Food Sci.* **2013**, *78*, 1752–1756. [CrossRef]
216. Duelge, K.; Nishshanka, U.; De Alwis, H. An LC-MS/MS method for the determination of antibiotic residues in distillers grains. *J. Chromatogr. B Analyt. Technol. Biomed. Life Sci.* **2017**, *1053*, 81–86. [CrossRef]
217. Mohanty, B.; Mahanty, A.; Ganguly, S.; Sankar, T.V.; Chakraborty, K.; Rangasamy, A.; Paul, B.; Sarma, D.; Mathew, S.; Kunnath Asha, K.; et al. Amino acid compositions of 27 food fishes and their importance in clinical nutrition. *J. Amino Acids* **2014**, *2014*, 269797. [CrossRef]
218. 218Ribeiro Alvarenga, R.; Borges Rodrigues, P.; de Souza Cantarelli, V.; Gilberto Zangeronimo, M.; da Silva Júnior, J.; da Silva, L.; Moreira dos Santos, L.; Pereira, L. Energy values and chemical composition of spirulina (*Spirulina platensis*) evaluated with broilers. *Rev. Bras. Zootec.* **2011**, *40*, 992–996. [CrossRef]
219. Campanella, L.; Russo, M.V.; Avino, P. Free and total amino acid composition in blue-green algae. *Ann. Chim.* **2002**, *92*, 343–352. [PubMed]
220. Abdullah Al-Dhabi, N.; Valan Arasu, M. Quantification of Phytochemical from Commercial *Spirulina* Products and Their Antioxidant Activities. *Evid. Based Complement. Alternat. Med.* **2016**, *2016*, 7631864.

221. Dziągwa-Becker, M.M.; Ramos, J.M.M.; Topolski, J.K.; Oleszek, W.A. Determination of free amino acids in plants by liquid chromatography coupled to tandem mass spectrometry (LC-MS/MS). *BioSci. Trends* **2011**, *5*, 231–238. [CrossRef]
222. Ma, X.; Zhao, D.; Li, X.; Meng, L. Chromatographic method for determination of the free amino acid content of chamomile flowers. *Pharmacogn. Mag.* **2015**, *11*, 176–179. [PubMed]
223. Salazar-Villanea, S.; Bruininx, E.M.A.M.; Gruppen, H.; Hendriks, W.H.; Carré, P.; Quinsac, A.; van der Poel, A.F.B. Physical and chemical changes of rapeseed meal proteins during toasting and their effects on in vitro digestibility. *J. Anim. Sci. Biotechnol.* **2016**, *7*, 62. [CrossRef]
224. Jajić, I.; Krstović, S.; Glamočić, D.; Jakšić, S.; Abramović, B. Validation of an HPLC method for the determination of amino acids in feed. *J. Serbian Chem. Soc.* **2013**, *78*, 839–850. [CrossRef]
225. Szkudzińska, K.; Smutniak, H.; Rubaj, J.; Korol, W.; Bielecka, G. Method validation for determination of amino acids in feed by UPLC. *Accredit. Qual. Assur.* **2018**, *22*, 247–252. [CrossRef]
226. Desmarais, S.M.; Cava, F.; de Pedro, M.A.; Casey Huang, K. Isolation and Preparation of Bacterial Cell Walls for Compositional Analysis by Ultra Performance Liquid Chromatography. *J. Vis. Exp.* **2014**, *83*, e51183. [CrossRef]
227. Kühner, D.; Stahl, M.; Demircioglu, D.D.; Bertsche, U. From cells to muropeptide structures in 24 h: Peptidoglycan mapping by UPLC-MS. *Sci. Rep.* **2014**, *4*, 7494. [CrossRef]
228. Marseglia, A.; Sforza, S.; Faccini, A.; Bencivenni, M.; Palla, G.; Caligian, A. Extraction, identification and semi-quantification of oligopeptides in cocoa beans. *Food Res. Int.* **2014**, *63*, 382–389. [CrossRef]
229. Prados, I.M.; Marina, M.L.; García, M.C. Isolation and identification by high resolution liquid chromatography tandem mass spectrometry of novel peptides with multifunctional lipid lowering capacity. *Food Res. Int.* **2018**, *111*, 77–86. [CrossRef] [PubMed]
230. Holman, B.W.B.; Malau-Aduli, A.E.O. Spirulina as livestock supplement and animal feed. *J. Anim. Physiol. Anim. Nutr.* **2013**, *97*, 615–623. [CrossRef] [PubMed]
231. Nurcahya Dewi, E.; Amalia, U.; Mel, M. The effect of Different Treatments to the Amino Acid Contents of Micro Algae *Spirulina* sp. *Aquatic Procedia* **2016**, *7*, 59–65. [CrossRef]
232. Wang, Y.; Shen, K.; Li, P.; Zhou, J.; Chao, Y. Simultaneous determination of 20 underivatized amino acids by high performance liquid chromatography-evaporative light-scattering detection. *Se Pu* **2011**, *29*, 908–911. [PubMed]
233. Prinsen, H.C.M.T.; Schiebergen–Bronkhorst, B.G.M.; Roeleveld, M.N.; Jans, J.J.M.; de Sain-van der Velden, M.G.M.; Visser, G.; van Hasselt, P.M.; Verhoeven-Duif, N.M. Rapid quantification of underivatized amino acids in plasma by hydrophilic interaction liquid chromatography (HILIC) coupled with tándem mass-spectrometry. *J. Inherit. Metab. Dis.* **2016**, *39*, 651–660. [CrossRef]
234. Masuda, A.; Dohmae, N. Amino acid analysis of sub-picomolar amounts of proteins by pre-column fluorescence derivatization with 6-aminoquinolyl-*N*-hydroxysuccinimidyl carbamate. *Biosc. Trends* **2011**, *5*, 231–238. [CrossRef]
235. Dhillon, M.K.; Kumar, S.; Gujar, G.T. A common HPLC-PDA method for amino acid analysis in insects and plants. *Indian J. Exp. Biol.* **2014**, *52*, 73–79.
236. Artavia, G.; Rojas-Bogantes, L.; Granados-Chinchilla, F. Two alternative chromatography methods assisted by the sulfonic acid moiety for the determination of furosine in milk. *MethodsX* **2018**, *5*, 639–647. [CrossRef]
237. Dolowy, M.; Pyka, A. Application of TLC, HPLC and GC methods to the study of amino acid and peptide enantiomers: A review. *Biomed. Chromatogr.* **2013**. [CrossRef]
238. Bartolomeo, M.; Maisano, F. Validation of a Reversed-Phase HPLC Method for Quantitative Amino Acid Analysis. *J. Biomol. Tech.* **2006**, *17*, 131–137.
239. Zhou, X.; Zhang, J.; Pan, Z.; Li, D. Review of Methods for the Detection and Determination of Malachite Green and Leuco-Malachite Green in Aquaculture. *Crit. Rev. Anal. Chem.* **2018**, *14*, 1–20. [CrossRef] [PubMed]
240. Adel, M.; Dadar, M.; Oliveri Conti, G. Antibiotics and malachite green in farmed rainbow trout (*Oncorhynchus mykiss*) from Iranian markets: A risk assessment. *Int. J. Food Prop.* **2017**, *20*, 402–408. [CrossRef]
241. Sudova, E.; Machova, J.; Svobodova, Z.; Vesely, T. Negative effects of malachite green and possibilities of its replacement in the treatment of fish eggs and fish: A review. *Vet. Med.* **2007**, *52*, 527–539. [CrossRef]
242. Hidayah, N.; Abu Bakar, F.; Mahyudin, N.A.; Faridah, S.; Nur-Azura, M.S.; Zaman, M.Z. Detection of malachite green and leuco-malachite green in fishery industry. *Int. Food Res. J.* **2013**, *20*, 1511–1519.

243. Xie, J.; Peng, T.; Chen, D.; Zhang, Q.; Wang, G.; Wang, X.; Guo, Q.; Jiang, F.; Chen, D.; Deng, J. Determination of malachite green, crystal violet and their leuco-metabolites in fish by HPLC-VIS detection after immunoaffinity column clean-up. *J. Chromatogr. B* **2013**, *913–914*, 123–128. [CrossRef] [PubMed]
244. Chen, G.; Miao, S. HPLC Determination and MS Confirmation of Malachite Green, Gentian Violet, and Their Leuco Metabolite Residues in Channel Catfish Muscle. *J. Agric. Food Chem.* **2010**, *58*, 7109–7114. [CrossRef] [PubMed]
245. Wang, Y.; Liao, K.; Huang, X.; Yuan, D. Simultaneous determination of malachite green, crystal violet and their leuco-metabolites in aquaculture water samples using monolithic fiber-based solid-phase microextraction coupled with high performance liquid chromatography. *Anal. Methods* **2015**, *7*, 8138. [CrossRef]
246. Bae Lee, J.; Yun Kim, H.; Mi Jang, Y.; Young Song, J.; Min Woo, S.; Sun Park, M.; Sook Lee, H.; Kyu Lee, S.; Kim, M. Determination of malachite green and crystal violet in processed fish products. *Food Addit. Contam. Part A* **2010**, *27*, 953–961. [CrossRef]
247. Chengyun, Z.; Jie, W.; Xuefang, D.; Zhimou, G.; Mingyang, L.; Xinmiao, L. Fast analysis of malachite green, leucomalachite green, crystal violet and leucocrystal violet in fish tissue based on a modified QuEChERS procedure. *Chin. J. Chromatogr.* **2014**, *4*, 419–425.
248. Turnipseed, S.B.; Andersen, W.C.; Roybal, J.E. Determination and Confirmation of Leucomalachite Green in Salmon using No-Discharge Atmospheric Pressure Chemical Ionization LC-MS. *J. AOAC Int.* **2005**, *88*, 1312–1317.
249. Abro, K.; Mahesar, S.A.; Iqbal, S.; Perveen, S. Quantification of malachite green in fish feed utilizing liquid chromatography-tandem mass spectrometry with a monolithic column. *Food Addit. Contam. Part A* **2014**, *31*, 827–832. [CrossRef] [PubMed]
250. EFSA. Malachite green in food, EFSA Panel on Contaminants in the Food Chain (CONTAM). *EFSA J.* **2016**, *14*, 4530. [CrossRef]
251. Furusawa, N. An isocratic Toxic Chemical-Free Mobile Phase HPLC-PDA analysis of Malachite Green and Leuco-Malachite Green. *Chromatography* **2014**, *1*, 75–81. [CrossRef]
252. Bilandžić, N.; Varenina, I.; Solomun Kolanović, B.; Oraić, D.; Zrnčić, S. Malachite green residues in farmed fish in Croatia. *Food Control* **2012**, *26*, 393–396. [CrossRef]
253. Barani, A.; Tajik, H. Malachite green residue in farmed fish in north-west part of Iran. *Int. J. Food Prop.* **2017**, *20*, S580–S585. [CrossRef]
254. Bedale, W.; Sindelar, J.J.; Milkowski, A.L. Dietary nitrate and nitrite: Benefits, risks, and evolving perceptions. *Meat Sci.* **2016**, *120*, 85–92. [CrossRef] [PubMed]
255. Wakida, F.T.; Lerner, D.N. Non-agricultural sources of groundwater nitrate: Review and case study. *Water Res.* **2005**, *39*, 3–16. [CrossRef]
256. Deutsche Forschungsgemeinschaft. *Nitrate and Nitrite in Diet: An Approach to Assess Benefit and Risk for Human Health*; Institut für Lebensmitteltoxikol: Hannover, Germany, 2014; p. 42.
257. Iammarino, M.; Di Taranto, A.; Cristino, M. Monitoring of nitrites and nitrates levels in leafy vegetables (spinach and lettuce): A contribution to risk assessment. *J. Sci. Food Agric.* **2014**, *94*, 773–778. [CrossRef]
258. Lundberg, J.O.; Gladwin, M.T.; Ahluwalia, A.; Benjamin, N.; Bryan, N.S.; Butler, A.; Cabrales, P.; Fago, A.; Feelisch, M.; Ford, P.C.; et al. Nitrate and nitrite in biology, nutrition and therapeutics. *Nat. Chem. Biol.* **2009**, *5*, 865–869. [CrossRef]
259. Shiva, S. Nitrite: A physiological store of nitric oxide and modulator of mitochondrial function. *Redox Biol.* **2013**, *1*, 40–44. [CrossRef]
260. Espejo-Herrera, N.; Gràcia-Lavedan, E.; Boldo, E.; Aragonés, N.; Pérez-Gómez, B.; Pollán, M.; Molina, A.J.; Fernández, T.; Martín, V.; La Vecchia, C.; et al. Colorectal cancer risk and nitrate exposure through drinking water and diet. *Int. J. Cancer* **2016**, *139*, 334–346. [CrossRef] [PubMed]
261. Ward, M.H. Too Much of Good Thing? Nitrate from Nitrogen Fertilizers and Cancer. *Rev. Environ. Health* **2009**, *24*, 357–363. [CrossRef] [PubMed]
262. European Commission. Regulation (EC) No 1881/2006 of 19 December 2006 setting levels for certain contaminants in foodstuff. *Off. J. Eur. Communities* **2006**, *016.011*, 1–35.
263. Brkić, D.; Bošnir, J.; Bevardi, M.; Gross Bošković, A.; Miloš, S.; Lasić, D.; Krivohlavek, A.; Racz, A.; Mojsović-Ćuić, A.; Uršulin Trstenjak, N. Nitrate in leafy green vegetables and estimated intake. *Afr. J. Tradit. Complement. Altern. Med.* **2017**, *14*, 31–41. [PubMed]

264. Pardo, O.; Yusà, V.; Villalba, P.; Perez, J.A. Monitoring programme on nitrate in vegetables and vegetable-based baby foods marketed in the Region of Valencia: Levels and estimated daily intake. *Food Addit. Contam.* **2010**, *27*, 478–486. [CrossRef] [PubMed]
265. Quijano, L.; Yusà, V.; Font, G.; McAllister, C.; Torres, C.; Pardo, O. Risk assessment and monitoring programme of nitrates through vegetables in the Region of Valencia (Spain). *Food Chem. Toxicol.* **2017**, *100*, 42–49. [CrossRef] [PubMed]
266. Hsu, J.; Arcot, J.; Lee, N.A. Nitrate and nitrite quantification from cured meat and vegetables and their estimated dietary intake in Australians. *Food Chem.* **2009**, *115*, 334–339. [CrossRef]
267. Dumitru Croitoru, M. Nitrite and nitrate can be accurately measured in samples of vegetal and animal origin using an HPLC-UV/VIS technique. *J. Chromatogr. B* **2012**, *911*, 154–161. [CrossRef]
268. Tsikas, D. Analysis of nitrite and nitrate I biological fluids by assays based on the Griess reaction: Appraisal of the Griess reaction in the L-arginine/nitric oxide area of research. *J. Chromatogr. B* **2007**, *851*, 51–70. [CrossRef]
269. Dumitru Croitoru, M.; Fülöp, I.; Miklos, A.; Hosszú, B.; Tátar, V.; Muntean, D. Presence of nitrate and nitrite in vegetables grown for self-consumption. *Farmacia* **2015**, *63*, 530–533.
270. Yagoub Abdulkair, B.; Elzupir, A.O.; Alamer, A.S. An Ultrasound Assessed Extraction Combined with Ion-Pair HPLC Method and Risk Assessment of Nitrite and Nitrate in Cured Meat. *J. Anal. Methods Chem.* **2018**, *2018*, 1907151. [CrossRef] [PubMed]
271. Chou, S.; Chung, J.; Hwang, D. A High Performance Liquid Chromatography Method for Determining Nitrate and Nitrite Levels in Vegetables. *J. Food Drug Anal.* **2003**, *11*, 233–238.
272. Nemade, K.; Fegade, U.; Ingle, S.; Attarde, S. High Performance Liquid Chromatography Method for Determination of Nitrite and Nitrate in Vegetable and Water samples. *Int. J. Adv. Sci. Tech. Res.* **2014**, *4*, 238–250.
273. Scheeren, M.B.; Arul, J.; Gariépy, C. Comparison of different method for nitrite and nitrate determination in meat products. In Proceedings of the 59th International Congress of Meat Science and Technology, Izmir, Turkey, 18–23 August 2013.
274. Dos Santos Baião, D.; Conte-Junior, C.; Flosi Paschoalin, V.; Silveira Alvares, T. Quantitative and Comparative Contents of Nitrate and Nitrite in *Beta vulgaris* L. by Reversed. Phase High-Performance Liquid Chromatography-Fluorescence. *Food Anal. Methods* **2016**, *9*, 1002–1008. [CrossRef]
275. Casanova, J.A.; Gross, L.K.; McMullen, S.E.; Schenck, F.J. Use of Greiss reagent containing Vanadium(III) fro post-column derivatization and simultaneous determination of nitrite and nitrate in baby food. *J. AOAC Int.* **2006**, *89*, 447–451. [PubMed]
276. Marcus, Y. Thermodynamics of Solvation of Ions. *J. Chem. Soc. Faraday Trans.* **1991**, *87*, 2995–2999. [CrossRef]
277. Oruc, H.H.; Akkoc, A.; Uzunoglu, I.; Kennerman, E. Nitrate Poisoning in Horses Associated with Ingestion of Forage and Alfalfa. *J. Equine Vet. Sci.* **2010**, *30*, 159–162. [CrossRef]
278. Merino, L.; Örnemark, U.; Toldrá, F. Chapter Three—Analysis of Nitrite and Nitrate in Foods: Overview of Chemical, Regulatory and Analytical Aspects. *Adv. Food Nutr. Res.* **2017**, *81*, 65–107. [PubMed]
279. Wang, Q.; Yu, L.; Liu, Y.; Lin, L.; Lu, R.; Zhu, J.; He, L.; Lu, Z. Methods for the detection and determination of nitrite and nitrate: A review. *Talanta* **2017**, *165*, 709–720. [CrossRef]
280. O'Neil, C.A.; Schwartz, S.J. Chromatographic analysis of cis/trans carotenoid isomers. *J. Chromatogr.* **1992**, *624*, 235–252. [CrossRef]
281. Wilberg, V.C.; Rodriguez-Amaya, D.B. HPLC Quantitation of Major Carotenoids of Fresh and Processed Guava, Mango and Papaya. *LWT-Food Sci. Technol.* **1995**, *28*, 474–480. [CrossRef]
282. Zanatta, C.F.; Mercadante, A.Z. Carotenoid composition from the Brazilian tropical fruit camu–camu (*Myrciaria dubia*). *Food Chem.* **2007**, *101*, 1526–1532. [CrossRef]
283. Chen, J.P.; Tai, C.Y.; Chen, B.H. Improved liquid chromatographic method for determination of carotenoids in Taiwanese mango (*Mangifera indica* L.). *J. Chromatogr. A* **2004**, *29*, 261–268. [CrossRef]
284. Inbaraj, B.S.; Chien, J.T.; Chen, B.H. Improved High Performance Liquid Chromatographic Method for Determination of Carotenoids in the Microalga Chlorella pyrenoidosa. *J. Chromatogr. A* **1102**, *1102*, 193–199. [CrossRef] [PubMed]
285. McGraw, K.J.; Toomey, M.B. Carotenoid accumulation in the tissues of zebra finches: Predictors of integumentary pigmentation and implications for carotenoid allocation strategies. *Physiol. Biochem. Zool.* **2010**, *83*, 97–109. [CrossRef] [PubMed]

286. Aluç, Y.; Kankılıç, G.B.; Tüzün, I. Determination of carotenoids in two algae species from the saline water of Kapulukaya reservoir by HPLC. *J. Liquid Chromatogr. Relat. Technol.* **2018**, *41*, 1–8. [CrossRef]
287. Shih-Chuan, L.; Jau-Tien, L.; Deng-Jye, Y. Determination of cis- and trans- α- and β-carotenoids in Taiwanese sweet potatoes (*Ipomoea batatas* (L.) Lam.) harvested at various times. *Food Chem.* **2009**, *116*, 605–610.
288. Van Jaarsveld, P.J.; Marais, D.W.; Harmse, E.; Nestel, P.; Rodriguez-Amaya, D.B. Retention of β-carotene in boiled, mashed orange-fleshed sweet potato. *J. Food Compos. Anal.* **2006**, *19*, 321–329. [CrossRef]
289. Huck, C.; Popp, M.; Scherz, H.; Bonn, G.K. Development and Evaluation of a New Method for the Determination of the Carotenoid Content in Selected Vegetables by HPLC and HPLC–MS–MS. *J. Chromatogr. Sci.* **2000**, *38*, 441–449. [CrossRef]
290. Lessin, W.J.; Catigani, G.L.; Schwartz, S.J. Quantification of cis-trans Isomers of Provitamin A Carotenoids in Fresh and Processed Fruits and Vegetables. *J. Agric. Food Chem.* **1997**, *45*, 3728–3732. [CrossRef]
291. Gayosso-García, L.E.; Yahia, E.M.; González-Aguila, G.A. Identification and quantification of phenols, carotenoids, and vitamin C from papaya (*Carica papaya* L., cv. Maradol) fruit determined by HPLC-DAD-MS/MS-ESI. *Food Res. Int.* **2011**, *44*, 1284–1291. [CrossRef]
292. Schweiggert RFSteingass, C.B.; Esquivel, P.; Carle, R. Chemical and Morphological Characterization of Costa Rican Papaya (*Carica papaya* L.) Hybrids and Lines with Particular Focus on Their Genuine Carotenoid Profiles. *J. Agric. Food Chem.* **2012**, *60*, 2577–2585.
293. Chacón-Ordóñez, T.; Schweiggert, R.M.; Bosy-Westphal, A.; Jiménez, V.M.; Carle, R.; Esquivel, P. Carotenoids and carotenoid esters of orange- and yellow-fleshed mamey sapote (*Pouteria sapota* (Jacq.) H.E. Moore & Stearn) fruit and their postprandial absorption in humans. *Food Chem.* **2017**, *221*, 673–682. [PubMed]
294. Rojas-Garbanzo, C.; Gleichenhagen, M.; Heller, A.; Esquivel, P.; Schulze-Kaysers, N.; Schieber, A. Carotenoid profile, antioxidant capacity, and chromoplasts of pink guava (*Psidium guajava* L. Cv. 'Criolla') during fruit ripening. *J. Agric. Food Chem.* **2017**, *65*, 3737–3747. [CrossRef] [PubMed]
295. Irias-Mata, A.; Jiménez, V.M.; Steingass, C.B.; Schweiggert, R.M.; Carle, R.; Esquivel, P. Carotenoids and xanthophyll esters of yellow and red nance fruits (*Byrsonima crassifolia* (L.) Kunth) from Costa Rica. *Food Res. Int.* **2018**, *111*, 708–714. [CrossRef] [PubMed]
296. Marutti, L.R.B.; Mercadante, A.Z. Carotenoid esters analysis and occurrence: What do we know so far? *Arch. Biochem. Biophys.* **2018**, *648*, 36–43. [CrossRef] [PubMed]
297. Wen, X.; Hempel, J.; Schweiggert, R.M.; Ni, Y.; Carle, R. Carotenoids and carotenoid esters of red and yellow *Physalis* (*Physalis alkekengi* L. and *P. pubescens* L.) fruits and calyces. *J. Agric. Food Chem.* **2017**, *65*, 6140–6151. [CrossRef] [PubMed]
298. De Rosso, V.; Mercadante, A. Identification and Quantification of Carotenoids, by HPLC-PDA-MS/MS, from Amazonian Fruits. *J. Agric. Food Chem.* **2007**, *55*, 5062–5072. [CrossRef]
299. Ligor, M.; Kováčová, J.; Gadzała-Kopciuch, R.; Studzińska, S.; Bocian, S.; Lehotay, J.; Buszewski, B. Study of RP HPLC Retention Behaviours in Analysis of Carotenoids. *Chromatographia* **2014**, *77*, 1047–1057. [CrossRef]
300. Schex, R.; Lieb, V.; Jiménez, V.M.; Esquivel, P.; Schweiggert, R.; Carle, R.; Steingrass, C.B. HPLC-DAD-APCI/ESI-MS^n analysis of carotenoids and α-tocopherol in Costa Rican *Acrocomia aculeata* fruits of varying maturity stages. *Food Res. Int.* **2018**, *105*, 645–653. [CrossRef]
301. Schweiggert, R.M.; Vargas, E.; Conrad, J.; Hempel, J.; Gras, C.C.; Ziegler, J.U.; Mayer, A.; Jiménez, V.; Esquivel, P.; Carle, R. Carotenoids, carotenoid esters, and anthocyanins of yellow-, orange-, and red-peeled cashew apples (*Anacardium occidentale* L.). *Food Chem.* **2016**, *200*, 274–282. [CrossRef] [PubMed]
302. Chacón-Ordoñez, T.; Esquivel, P.; Jiménez, V.M.; Carle, R.; Schweiggert, R.M. Deposition Form and Bioaccessibility of Keto-Carotenoids from Mamey Sapote (*Pouteria sapota*), Red Bell Pepper (*Capsicum annuum*), and Sockeye Salmon (*Oncorhynchus nerka*) Filet. *J. Agric. Food Chem.* **2016**, *64*, 1989–1998. [CrossRef] [PubMed]
303. Dolan, J.W. How Much Can I Inject? Part II: Injecting in Solvents Other than Mobile Phase. *LCGC N. Am.* **2014**, *32*, 854–859.
304. Pereira da Costa, M.; Conte-Junior, C.A. Chromatographic methods for the determination of carbohydrates and organic acids in foods of animal origin. *Compr. Rev. Food Sci. Food Saf.* **2015**, *14*, 586–600. [CrossRef]
305. De Goeij, S. Quantitative Analysis Methods for Sugars. Master's Thesis, Universiteit van Amsterdam, Amsterdam, The Netherlands, August 2013.

306. Agius, C.; von Tucher, S.; Poppenberger, B.; Rozhon, W. Quantification of sugars and organic acids in tomato fruits. *MethodsX* **2018**, *5*, 2537–2550. [CrossRef] [PubMed]
307. Xu, W.; Liang, L.; Zhu, M. Determination of Sugars in Molasses by HPLC Following Solid-Phase Extraction. *Int. J. Food Prop.* **2015**, *18*, 547–557. [CrossRef]
308. Koh, D.-W.; Park, J.-W.; Lim, J.-H.; Yea, M.-J.; Bang, D.-Y. A rapid method for simultaneous quantification of 13 sugars and sugar alcohols in food products by UPLC-ELSD. *Food Chem.* **2018**, *240*, 694–700. [CrossRef]
309. Zielinkski, A.A.F.; Braga, C.M.; Demiate, I.M.; Beltrame, F.L.; Nogueira, A.; Wosiaki, G. Development and optimization of a HPLC-RI method for the determination of major sugars in apple juice and evaluation of the effect of the ripening stage. *Food Sci. Technol.* **2014**, *34*, 38–43. [CrossRef]
310. Duarte-Delgado, D.; Narváez-Cuenca, C.E.; Restrepo-Sánchez, L.P.; Kushalappa, A.; Mosquera-Vásquez, T. Development and validation of a liquid chromatographic method to quantify sucrose, glucose, and fructose in tubers of *Solanum tuberosum* Group Phureja. *J. Chromatogr. B* **2015**, *975*, 18–23. [CrossRef]
311. Shindo, T.; Sadamasu, Y.; Suzuki, K.; Tanaka, Y.; Togawa, A.; Uematsu, Y. Method of quantitative analysis by HPLC and confirmation by LC-MS of sugar alcohols in foods. *Shokuhin Eiseigaku Zasshi* **2013**, *54*, 358–363. [CrossRef]
312. Canesin, R.C.F.S.; Isique, W.D.; Buzetti, S.; Aparecida de Souza, J. Derivation method for determining sorbitol in fruit trees. *Am. J. Plant Sci.* **2014**, *5*, 3457–3463. [CrossRef]
313. Hung, W.-T.; Chen, Y.-T.; Wang, S.-H.; Liu, Y.-C.; Yang, W.-B. A new method for aldo-sugar analysis in beverages and dietary foods. *Funct. Food Health Dis.* **2016**, *6*, 234–245.
314. Valliydan, B.; Shi, H.; Nguyen, H.T. A simple analytical method for high-throughput screening of major sugars from soybean by normal-phase hplc with evaporative light scattering detection. *Chromatogr. Res. Int.* **2015**, *2015*, 757649. [CrossRef]
315. Wieß, K.; Alt, M. Determination of single sugars, including inulin, in plants and feed materials by high-performance liquid chromatography and refraction index detection. *Fermentation* **2017**, *3*, 36.
316. Verspreet, J.; Pollet, A.; Cuyvers, S.; Vergauwen, R.; Van den Ende, W.; Delcour, J.A.; Courtin, C.M. A simple and accurate method for determining wheat grain fructan content and average degree of polymerization. *J. Agric. Food Chem.* **2012**, *60*, 2102–2107. [CrossRef] [PubMed]
317. Correia, D.M.; Dias, L.G.; Veloso, A.C.A.; Dias, T.; Rocha, I.; Rodrigues, L.R.; Peres, A.M. Dietary sugars analysis: Quantification of fructooligosaccharides during fermentation by HPLC-RI method. *Front. Nutr.* **2014**, *1*, 11. [CrossRef]
318. Salazar Murillo, M.M.; Granados-Chinchilla, F. Total starch in animal feeds and silages based on the chromatographic determination of glucose. *MethodsX* **2018**, *5*, 83–89. [CrossRef]
319. Bai, W.; Fang, X.; Zhao, W.; Huang, S.; Zhang, H.; Qian, M. Determination of oligosaccharides and monosaccharides in Hakka rice wine by precolumn derivation high-performance liquid chromatography. *J. Food Drug Anal.* **2015**, *23*, 645–651. [CrossRef]
320. Madhuri, K.V.; Prabhakar, K.V. Recent trends in the characterization of microbial exopolysaccharides. *Orient. J. Chem.* **2014**, *30*, 895–904. [CrossRef]
321. Corradini, C.; Cavazza, A.; Bignardi, C. High-performance anion-exchange chromatography coupled with pulsed electrochemical detection as a powerful tool to evaluate carbohydrates of food interest: Principles and applications. *Int. J. Carbohydr. Chem.* **2012**, *2012*, 487564. [CrossRef]
322. Dvořáčková, E.; Šnóblová, M.; Hrdlička, P. Carbohydrate analysis: From sample preparation to HPLC on different stationary phases coupled with evaporative light-scattering detection. *J. Sep. Sci.* **2014**, *37*, 323–337. [CrossRef] [PubMed]
323. Rippe, J.M.; Angelopoulos, T.J. Added sugars and risk factors for obesity, diabetes and heart disease. *Int. J. Obes.* **2016**, *40*, S22–S27. [CrossRef] [PubMed]
324. Louie, J.C.Y.; Moshtaghian, H.; Boylan, S.; Flood, V.M.; Rangan, A.M.; Barclay, A.W.; Brand-Miller, J.C.; Gill, T.P. A systematic methodology to estimate added sugar content of foods. *Eur. J. Clin. Nutr.* **2015**, *69*, 154–161. [CrossRef] [PubMed]
325. Uçar, G.; Balaban, M. Hydrolysis of polysaccharides with 77% sulfuric acid for quantitative saccharification. *Turk. J. Agric. For.* **2003**, *27*, 361–365.
326. Yan, X. High performance liquid chromatography for carbohydrate analysis. In *High-Performance Liquid Chromatography (HPLC): Principles, Practices and Procedures*; Zhuo, Y., Ed.; Nova Science Publishers: Hauppauge, NY, USA, 2014; ISBN 978-1-62948-854-7.

327. Thøgersen, R.; Castro-Mejía, J.L.; Sundekilde, U.K.; Hansen, L.H.; Hansen, A.K.; Nielsen, D.S.; Bertram, H.C. Ingestion of an inulin-enriched pork sausage product positively modulates the gut microbiome and metabolome of healthy rats. *Mol. Nutr. Food Res.* **2018**, *13*, e1800608. [CrossRef]
328. Szpylka, J.; Thiex, N.; Acevedo, B.; Albizu, A.; Angrish, P.; Austin, S.; Bach Knudsen, K.E.; Barber, C.A.; Berg, D.; Bhandari, S.D.; et al. Standard Method Performance Requirements (SMPRs®) 2018.002: Fructans in Animal Food (Animal Feed, Pet Food, and Ingredients). *J. AOAC Int.* **2018**, *101*, 1283–1284. [CrossRef] [PubMed]
329. Longland, A.C.; Byrd, B.M. Pasture nonstructural carbohydrates and equine laminitis. *J. Nut.* **2006**, *136*, 2099S–2102S. [CrossRef]
330. Stöber, P.; Bénet, S.; Hischenhuber, C. Simplified Enzymatic High-Performance Anion Exchange Chromatographic Determination of Total Fructans in Food and Pet Foods-Limitations and Measurement Uncertainty. *J. Agric. Food Chem.* **2004**, *52*, 2137–2146. [CrossRef]
331. Rühmann, B.; Schmid, J.; Sieber, V. High throughput exopolysaccharide screening platform: From strain cultivation to monosaccharide composition and carbohydrate fingerprinting in one day. *Carbohydr. Polym.* **2015**, *122*, 212–220. [CrossRef]
332. Affonso, R.C.L.; Voytena, A.P.L.; Fanan, S.; Pitz, H.; Coelho, D.S.; Hortmann, A.L.; Pereira, A.; Uarrota, V.G.; Hillmann, M.C.; Varela, L.A.C.; et al. Phytochemical composition, antioxidant activity, and the effect of the aqueous extract of coffee (*Coffea arabica* L.) bean residual press cake on the skin wound healing. *Oxidative Med. Cell. Longev.* **2016**, *2016*, 1923754. [CrossRef]
333. Shao, J.; Cao, W.; Qu, H.; Pan, J.; Gong, X. A novel quality by design approach for developing an HPLC method to analyze herbal extracts: A case study of sugar content analysis. *PLoS ONE* **2018**, *13*, e019515. [CrossRef] [PubMed]
334. Swetwiwathana, A.; Visessanguan, W. Potential of bacteriocin-producing lactic acid bacteria for safety improvements of traditional Thai fermented meat and human health. *Meat Sci.* **2015**, *109*, 101–105. [CrossRef] [PubMed]
335. Gurtler, J.B.; Mai, T.L. Traditional preservatives: Organic acids. *Encycl. Food Microbiol.* **2014**, *3*, 119–130.
336. Anyasi, T.A.; Jideani, A.I.O.; Edokpayi, J.N.; Anokwuru, C.P. Application of Organic Acids in Food Preservation. In *Organic Acids, Characteristics, Properties and Synthesis*; Vargas, C., Ed.; Nova Sciences Publishers, Inc.: New York, NY, USA, 2016; pp. 1–47. ISBN 9781634859523.
337. Neffe-Skocińska, K.; Sionek, B.; Ścibisz, I.; Kotoźyn-Krajewska, D. Acid contents and the effect of fermentation condition of Kombucha tea beverages on physicochemical, microbiological and sensory properties. *CyTA-J. Food* **2017**, *15*. [CrossRef]
338. Nour, V.; Trandafir, I.; Ionica, M.E. HPLC Organic Acid Analysis in Different Citrus Juices under Reversed Phase Conditions. *Not. Bot. Hort. Agrobot. Cluj* **2010**, *38*, 44–48.
339. Lobo Roriz, C.; Barros, L.; Carvalho, A.M.; Ferreira, I.C.F.R. HPLC-Profiles of Tocopherols, Sugars, and Organic Acids in Three Medicinal Plants Consumed as Infusions. *Int. J. Food Sci.* **2014**, *2014*, 241481. [CrossRef]
340. Scherer, R.; Rybka, A.C.P.; Ballus, C.A.; Dillenburg Meinhart, A.; Texeira Filho, J.; Texeira Godoy, H. Validation of HPLC method for simultaneous determination of main organic acids in fruits and juices. *Food Chem.* **2012**, *135*, 150–154. [CrossRef]
341. Llano, T.; Quijorna, N.; Andrés, A.; Coz, A. Sugar, acid and furfural quantification in a sulphite pulp mill: Feedstock, product and hydrolysate analysis by HPLC/RID. *Biotechnol. Rep.* **2017**, *15*, 75–83. [CrossRef]
342. Jain, I.; Kumar, V.; Satyanarayana, T. Xylooligosaccharides: An economical prebiotic from agroresidues and their health benefits. *Indian J. Exp. Biol.* **2015**, *53*, 131–142.
343. Saleh Zaky, A.; Pensupa, N.; Andrade-Eiroa, Á.; Tucker, G.A.; Du, C. A new HPLC method for simultaneously measuring chloride, sugars, organic acids and alcohols in food samples. *J. Food Compos. Anal.* **2017**, *56*, 25–33. [CrossRef]
344. Ergönül, P.G.; Nergiz, C. Determination of Organic Acids in Olive Fruit by HPLC. *Czech J. Food Sci.* **2010**, *28*, 202–205. [CrossRef]
345. Tašev, K.; Stefova, M.; Ivanova-Petropulos, V. HPLC method validation and application for organic acid analysis in wine after solid-phase extraction. *Maced. J. Chem. Chem. Eng.* **2016**, *35*. [CrossRef]
346. Mihaljević Žulj, M.; Puhelek, I.; Jagatić Korenika, A.M.; Maslov Bandić, L.; Pavlešić, T.; Jeromel, A. Organic Acid Composition in Croatian Predicate Wines. *Agric. Conspec. Sci.* **2015**, *80*, 113–117.

347. Sánchez-Machado, D.I.; López-Cervantes, J.; Martínez-Cruz, O. Quantification of Organic Acids in Fermented Shrimp Waste by HPLC. *Food Technol. Biotechnol.* **2008**, *46*, 456–460.
348. Ke, W.C.; Ding, W.R.; Xu, D.M.; Ding, L.M.; Zhang, P.; Li, F.D.; Guo, X.S. Effects of addition of malic or citric acids on fermentation quality and chemical characteristics of alfalfa silage. *J. Dairy Sci.* **2017**, *100*, 1–9. [CrossRef] [PubMed]
349. Fasciano, J.M.; Mansour, F.R.; Danielson, N.D. Ion-Exclusion High-Performance Liquid Chromatography of Aliphatic Organic Acids Using a Surfactant-Modified C18 Column. *J. Chromatogr. Sci.* **2016**, *54*, 958–970. [CrossRef]
350. Wang, Y.; Wang, J.; Cheng, W.; Zhao, Z.; Cao, J. HPLC method for the simultaneous quantification of the major organic acids in Angeleno plum fruit. *IOP Conf. Ser. Mater. Sci. Eng.* **2014**, *62*, 012035. [CrossRef]
351. Coblentz, W.K.; Akins, M.S. Silage review: Recent advances and future technologies for baled silages. *J. Dairy Sci.* **2018**, *101*, 4075–4092. [CrossRef] [PubMed]
352. Khan, N.A.; Yu, P.; Ali, M.; Cone, J.W.; Hendriks, W.H. Nutritive value of maize silage in relation to dairy cow performance and milk quality. *J. Sci. Food Agric.* **2015**, *95*, 238–258. [CrossRef]
353. Silva, V.P.; Pereira, O.G.; Leandro, E.S.; Da Silva, K.G.; Ribeiro, K.G.; Mantovani, H.C.; Santos, S.A. Effects of lactic acid bacteria with bacteriocinogenic potential on the fermentation profile and chemical composition of alfalfa silage in tropical conditions. *J. Dairy Sci.* **2016**, *99*, 1–8. [CrossRef] [PubMed]
354. Kim, H.J.; Lee, M.J.; Kim, H.J.; Cho, S.K. Development of HPLC-UV method for detection and quantification of eight organic acids in animal feed. *J. Chromatogr. Sep. Tech.* **2017**, *8*, 385. [CrossRef]
355. Grune, T.; Lietz, G.; Palou, A.; Ross, A.C.; Stahl, W.; Tang, G.; Thurnham, D.; Yin, S.A.; Biesalski, H.K. β-Carotene Is an Important Vitamin A Source for Humans. *J. Nutr.* **2010**, *140*, 2268S–2285S. [CrossRef] [PubMed]
356. Gilbert, C. What is vitamin A and why do we need it? *Community Eye Health* **2013**, *26*, 65. [PubMed]
357. Ding, Z.; Saldeen, T.G.P.; Mathur, P.; Mehta, J.L. Mixed tocopherols are better than alpha-tocopherol as antioxidant as good as statins. *Curr. Res. Cardiol.* **2016**, *3*, 128–129. [CrossRef]
358. Alshahrani, F.; Aljohani, N. Vitamin D: Deficiency, Sufficiency and Toxicity. *Nutrients* **2013**, *5*, 3605–3616. [CrossRef] [PubMed]
359. Ortiz-Boyer, F.; Fernandez-Romero, J.M.; Luque de Castro, M.D.; Quesada, J.M. Determination of vitamins D_2, D_3, K_1 and K_3 and some hydroxy metabolites of vitamin D_3 in plasma using a continuous clean-up–preconcentration procedure coupled on-line with liquid chromatography–UV detection. *Analyst* **1999**, *124*, 401–406. [CrossRef] [PubMed]
360. Schwalfenberg, G.K. Vitamins K1 and K2: The emerging group of vitamins required for human health. *J. Nutr. Metab.* **2017**, *345*, 229–234. [CrossRef] [PubMed]
361. Harshman, S.G.; Finnan, E.G.; Barger, K.J.; Bailey, R.L.; Haytowitz, D.B.; Gilhooly, C.H.; Booth, S.L. Vegetables and Mixed Dishes Are Top Contributors to Phylloquinone Intake in US Adults: Data from the 2011–2012 NHANES. *J. Nutr.* **2017**, *96*, 149–154. [CrossRef]
362. Beulens, J.; Booth, S.; van den Heuvel, E.; Stoecklin, E.; Baka, A.; Vermeer, C. The role of menaquinones (vitamin K2) in human health. *Br. J. Nutr.* **2013**, *110*, 1357–1368. [CrossRef]
363. Chavez-Servín, J.; Castellote, A.; Lopez-Sabater, M.C. Simultaneous analysis of Vitamins A and E in infant milk-based formulae by normal-phase high-performance liquid chromatography-diode array detection using a short narrow-bore column. *J. Chromatogr. A* **2003**, *1122*, 138–143. [CrossRef]
364. Płonka, J.; Toczek, A.; Tomczyk, V. Multivitamin Analysis of Fruits, Fruit–Vegetable Juices, and Diet Supplements. *Food Anal. Methods* **2012**, *5*, 1167–1176. [CrossRef]
365. Sami, R.; Li, Y.; Qi, B.; Wang, S.; Zhang, Q.; Han, F.; Ma, Y.; Jing, J.; Jiang, L. HPLC Analysis of Water-Soluble Vitamins (B2, B3, B6, B12, and C) and Fat-Soluble Vitamins (E, K, D, A, and β-Carotene) of Okra (*Abelmoschus esculentus*). *J. Chem.* **2014**, *2014*, 831357. [CrossRef]
366. Xue, X.; You, J.; He, P. Simultaneous Determination of Five Fat-Soluble Vitamins in Feed by High-Performance Liquid Chromatography Following Solid-Phase Extraction. *J. Chromatogr. Sci.* **2012**, *46*, 345–350. [CrossRef]
367. Qian, H.; Sheng, M. Simultaneous determination of fat-soluble vitamins A, D and E and pro-vitamin D2 in animal feeds by one-step extraction and high-performance liquid chromatography analysis. *J. Chromatogr. A* **1998**, *825*, 127–133. [CrossRef]

368. Lee, H.; Kwak, B.; Ahn, J.; Jeong, S.; Shim, S.; Kim, K.; Yoon, T.; Leem, D.; Jeong, J. Simultaneous Determination of Vitamin A and E in Infant Formula by HPLC with Photodiode Array Detection. *Korean J. Food Sci. Anim. Resour.* **2011**, *31*, 191–199. [CrossRef]
369. Lim, H.; Woo, S.; Sig Kim, H.; Jong, S.; Lee, J. Comparison of extraction methods for determining tocopherols in soybeans. *Eur. J. Lipid Sci. Technol.* **2007**, *109*, 1124–1127. [CrossRef]
370. Odes, S.; Hisil, Y. Analysis of vitamin A in eggs by high pressure liquid chromatography. *Die Nahr.* **1991**, *35*, 391–394.
371. Reynolds, S.; Judd, H. Rapid procedure for the determination of vitamins A and D in fortified skimmed milk powder using high-performance liquid chromatography. *Analyst* **1989**, *109*, 489–492. [CrossRef]
372. Turner, C.; Persson, M.; Mathiasson, L.; Adlercreutz, P.; King, J.W. Lipase-catalyzed reactions in organic and supercritical solvents: Application to fat-soluble vitamin determination in milk powder and infant formula. *Enzym. Microb. Technol.* **2001**, *29*, 111–121. [CrossRef]
373. Jedlička, A.; Klimeš, J. Determination of Water- and Fat-Soluble Vitamins in Different Matrices Using High-Performance Liquid Chromatography. *Chem. Papers* **2005**, *59*, 202–222. [CrossRef]
374. Rupérez, F.J.; Martín, D.; Herrera, E.; Barbas, C. Chromatographic analysis of alpha-tocopherol and related compounds in various matrices. *J. Chromatogr. A* **2001**, *935*, 45–69. [CrossRef]
375. Stefanov, I.; Vlaeminck, B.; Fievez, V. A novel procedure for routine milk fat extraction based on dichloromethane. *J. Food Compos. Anal.* **2010**, *23*, 852–855. [CrossRef]
376. Hung, G.W.C. Determination of Vitamins D2 and D3 in Feedingstuffs by High Performance Liquid Chromatography. *J. Liquid Chromatogr.* **1988**, *11*, 953–969. [CrossRef]
377. Lee, S.; Morrisa, Y.; Grossa, L.; Portera, F.; Ortiz-Colona, F.; Farrowa, J.; Waqara, A.; Phifera, E.; Kerdahia, K. Development and Single-Lab Validation of an UHPLC-APCI-MS/MS Method for Vitamin K1 in Infant Formulas and Other Nutritional Formulas. *J. Regul. Sci.* **2015**, *2*, 27–35.
378. Wagner, D.; Rousseau, D.; Sidhom, G.; Pouliot MAudet, T.; Vieth, R. Vitamin D3 Fortification, Quantification, and Long-Term Stability in Cheddar and Low-Fat Cheeses. *J. Agric. Food Chem.* **2008**, *56*, 7964–7969. [CrossRef]
379. Kienen, V.; Costa, W.; Visentainer, J.; Souza, N.; Oliveira, C. Development of a green chromatographic method for determination of fat-soluble vitamins in food and pharmaceutical supplement. *Talanta* **2008**, *75*, 141–146. [CrossRef] [PubMed]
380. Gomis, D.B.; Fernández, M.P.; Gutiérrez, M.D. Simultaneous determination of fat-soluble vitamins and provitamins in milk by microcolumn liquid chromatography. *J. Chromatogr. A* **2000**, *891*, 109–114. [CrossRef]
381. Koivu, T.; Piironen, V.; Henttonen, S.; Mattila, P. Determination of Phylloquinone in Vegetables, Fruits, and Berries by High-Performance Liquid Chromatography with Electrochemical Detection. *J. Chromatogr. A* **1997**, *45*, 4644–4649. [CrossRef]
382. Shammugasamy, B.; Ramakrishnan, Y.; Ghazali, H.; Muhammad, K. Tocopherol and tocotrienol contents of different varieties of rice in Malaysia. *J. Sci. Food Agric.* **2015**, *95*, 672–678. [CrossRef]
383. Kim, H.O. Development of an ion-pairing reagent and HPLC-UV method for the detection and quantification of six water-soluble vitamins in animal feed. *Int. J. Anal. Chem.* **2016**, *2016*, 8357358. [CrossRef]
384. Midttun, Ø.; McCann, A.; Aarseth, O.; Kokeide, M.; Kvalheim, G.; Meyer, K.; Ueland, P.M. Combined measurement of 6 fat-soluble vitamins and 26 water-soluble functional vitamin markers and amino acids in 50 μL of serum or plasma by high-throughput mass spectrometry. *Anal. Chem.* **2016**, *88*, 10427–10436. [CrossRef] [PubMed]
385. Kakitani, A.; Inoue, T.; Matsumoto, K.; Watanabe, J.; Nagatomi, Y.; Mochizuki, N. Simultaneous determination of water-soluble vitamins in beverages and dietary supplements by LC-MS/MS. *Food Addit. Contam. Part A* **2014**, *31*, 1939–1948. [CrossRef] [PubMed]
386. Zhang, Y.; Zhou, W.-E.; Yan, J.-Q.; Liu, M.; Zhou, Y.; Shen, X.; Ma, Y.-L.; Feng, X.-S.; Yang, J.; Li, G.-H. A Review of the Extraction and Determination Methods of Thirteen Essential Vitamins to the Human Body: An Update from 2010. *Molecules* **2018**, *23*, 1484. [CrossRef]
387. Abano, E.E.; Dadzie, R.G. Simultaneous detection of water-soluble vitamins using the High Performance Liquid Chromatography (HPLC)—A review. *Croat. J. Food Sci. Technol.* **2014**, *6*, 116–123. [CrossRef]

 © 2018 by the authors. Licensee MDPI, Basel, Switzerland. This article is an open access article distributed under the terms and conditions of the Creative Commons Attribution (CC BY) license (http://creativecommons.org/licenses/by/4.0/).

Article

Native Colombian Fruits and Their by-Products: Phenolic Profile, Antioxidant Activity and Hypoglycaemic Potential

Monica Rosa Loizzo [1], Paolo Lucci [2,*], Oscar Núñez [3], Rosa Tundis [1], Michele Balzano [4], Natale Giuseppe Frega [4], Lanfranco Conte [2], Sabrina Moret [2], Daria Filatova [3], Encarnación Moyano [3] and Deborah Pacetti [4]

1. Department of Pharmacy, Health and Nutritional Sciences, University of Calabria, 87036 Rende (CS), Italy; monica_rosa.loizzo@unical.it (M.R.L.); rosa.tundis@unical.it (R.T.)
2. Department of Agri-Food, Animal and Environmental Sciences, University of Udine, via Sondrio 2/a, 33100 Udine, Italy; lanfranco.conte@uniud.it (L.C.); sabrina.moret@uniud.it (S.M.)
3. Department of Chemical Engineering and Analytical Chemistry, University of Barcelona, Martí i Franquès 1-11, 08028 Barcelona, Spain; oscar.nunez@ub.edu (O.N.); daria.filatova@gmail.com (D.F.); encarna.moyano@ub.edu (E.M.)
4. Department of Agricultural, Food, and Environmental Sciences, Marche Polytechnic University, Via Brecce Bianche, 60131 Ancona, Italy; m.balzano@staff.univpm.it (M.B.); n.g.frega@univpm.it (N.G.F.); d.pacetti@staff.univpm.it (D.P.)
* Correspondence: paolo.lucci@uniud.it; Tel.: +39-0432-558170

Received: 25 January 2019; Accepted: 27 February 2019; Published: 3 March 2019

Abstract: The phenols and fatty acids profile and in vitro antioxidant and hypoglycaemic activity of seed, peel, pulp or pulp plus seeds of Colombian fruits from Solanaceae and Passifloraceae families were investigated. Ultra-High Performance Liquid Chromatography (UHPLC)-High Resolution Mass Spectrometry (HRMS) revealed the presence of chlorogenic acid as dominant phenolic compound in Solanaceae samples. Based on the Relative Antioxidant Score (RACI) and Global Antioxidant Score (GAS) values, *Solanum quitoense* peel showed the highest antioxidant potential among Solanaceae samples while *Passiflora tripartita* fruits exhibited the highest antioxidant effects among Passifloraceae samples. *P. ligularis* seeds were the most active as hypoglycaemic agent with IC_{50} values of 22.6 and 24.8 µg/mL against α-amylase and α-glucosidase, respectively. Considering that some of the most promising results were obtained by the processing waste portion, its use as functional ingredients should be considered for the development of nutraceutical products intended for patients with disturbance of glucose metabolism.

Keywords: Passifloraceae; Solanaceae; liquid chromatography; hypoglycaemic; α-amylase

1. Introduction

Colombia is a country with high levels of biodiversity and is home to a wide variety of exotic fruits. Traditionally, tropical fruit are consumed locally; nowadays, increasing production and more efficient transportation and refrigeration systems have led to increased global consumption with considerable quantities of tropical fruits that are now exported on a global scale. Tropical fruits not only present an attractive and characteristic exotic taste and aroma, but also represent a valuable source of bioactive compounds beneficial for humans. However, given the diversity of Colombian native and exotic species, the health properties and the chemical composition of some of these fruits have not been sufficiently studied.

It is well known that fruit phytochemicals play an important role against many ailments, including heart disease, diabetes, high blood pressure, cancer, etc. Among them, diabetes mellitus (DM) is projected to reach pandemic proportion in the next 25 years [1]. The persistent hyperglycaemia status that characterises both forms of DM causes an increase in the production of reactive oxygen species (ROS) both of cytosolic and mitochondrial origin. ROS could determine the long-term deterioration of pancreatic islet β-cell by affecting mitochondrial Adenosine triphosphate (ATP) production that is necessary for insulin secretion. The consequent mitochondrial dysfunction influences insulin sensitivity within muscle, liver and adipose tissue [2]. One of the most common approaches to reduce the intake of sugar is by the reduction of their intestinal absorption using carbohydrates-hydrolysing enzymes inhibitors. In our previous research articles, the potential role of natural products as α-amylase and α-glucosidase inhibitors and antioxidant compounds was demonstrated [3–5].

In this context, to enhance scientific knowledge for the composition and health benefits of Colombian fruits and their by-products, we selected *Solanum quitoense*, *Physalis peruviana*, and *Cyphomandra betacea* (synonymous *Solanum betaceum* Cav) belonging to Solanaceae family, and *Passiflora pinnatistipula*, *Passiflora tripartita* and *Passiflora ligularis* belonging to Passifloraceae family.

S. quitoense has a characteristic *Citrus* flavour, sometimes described as a combination of rhubarb and lime [6]. Fruits are used to prepare juice or a drink called lulada. The fruit of *P. peruviana* is a smooth berry largely used to prepare snacks, pies or jams. It is used also in salads combined with avocado [7]. *Cyphomandra betacea* is one of the most popular fruit in South America where it is used to make juice. Fruits of *P. tripartita* are generally consumed raw or as ingredient of ice creams, fruit salads, pies, jellies or to make drinks [7]. *P. ligularis* fruit is also employed for the preparation of processed products like marmalade and jelly [3], while *P. pinnastipula* is consumed as drinks, ice-cream or marmalades [7].

After the selection, we have screened the fatty acids and the phenolics profile, as well as the antioxidant and hypoglycaemic potential of extracts from peel, pulp, seed or pulp plus seed of selected Colombian fruits in order to investigate the functional potential of each fruits portion, and to identify a potential use of the processing waste of these extracts as functional ingredients.

2. Materials and Methods

2.1. Chemicals and Reagents

All chemicals and reagents used in this study were purchased from Sigma-Aldrich Chemical Co. Ltd. (Milan, Italy) and VWR International (Milan, Italy) and, unless specified otherwise, were analytical grade or higher.

2.2. Plant Material and Extraction Procedure

Fruits were purchased at the local market in Bogotá (Colombia). At least 10 fruits were combined for each of the three replicated samples (one per month). Samples were examined and cleaned by using distilled water. Pulp, peel and seed were manually separated wherever it is possible in order to submit each portion to extraction procedure. The freeze-dried vegetable material (2.5 g of clean seeds or 5 g of peel or 5 g of pulp or seed + pulp) were finely ground with an Ultraturrax device (IKA®-Werke GmbH & Co. KG, Königswinter, Germany) and undergone to cold extraction with absolute ethanol (40 mL) for 24 h at room temperature (20 °C). The top phase was filtered through Whatman filter paper #4 (9.0 cm diameter) and the residue was re-extracted with 30 mL of absolute ethanol. The filtered top phases were combined and dried in a rotary evaporator (Buchi R-300, Milan, Italy) with operating temperature of 35 °C. Extraction yield ranged from 7.2 to 36.5 g ethanolic extracts on 100 g dry product in Solanaceae family and from 2.2 to 49.1 g/100 g dry product in Passifloraceae family. Extracts were then stored at −20 °C until further analysis.

2.3. Ultra-High Performance Liquid Chromatography (UHPLC)-High Resolution Mass Spectrometry (HRMS). Conditions and Analysis of Phenolic Compounds in Colombian Fruits

UHPLC-HRMS analysis was carried out using an Accela UHPLC system (Thermo Fisher Scientific, San Jose, CA, USA) equipped with a quaternary pump. For the chromatographic separation, an Ascentis Express C18 (150 × 2.1 mm, 2.7 µm, Supelco) column was employed. The mobile phase was a mixture of solvent A (0.1% formic acid in water) and solvent B (0.1% formic acid in acetonitrile) and the flow rate was 300 µL/min. Gradient elution program was run as stated: from 0–1 min, 10% B; 1–20 min, linear gradient from 10 to 95% B; 20–23 min, isocratic step at 95% B; 23–25 min, back to initial conditions at 10% B, and 25–30 min, isocratic step at 10% B for column re-equilibration. The UHPLC system was coupled to a Q-Exactive quadrupole-Orbitrap (Thermo Fisher Scientific) mass spectrometer equipped with a heated electrospray probe ionization source (H-ESI II). HESI-II was operated in negative ionisation mode. Nitrogen was used as a sheath gas, sweep gas and auxiliary gas at flow rates of 60, 0 and 10 a.u. (arbitrary units), respectively. Heater temperature was set at 350 °C. Capillary temperature was set at 320 °C and electrospray voltage at −2.5 kV. The HRMS instrument was operated in full MS scan with a m/z range from 100 to 1500, and the mass resolution tuned into 70,000 full width half maximum (FWHM) at m/z 200, with an automatic gain control (AGC) target (the number of ions to fill C-Trap) of 5.0e5 with a maximum injection time (IT) of 200 ms. Selected analytes belonging to different phenolic classes, namely gallic acid, (+)-catechin hydrate, p-coumaric acid, chlorogenic acid, (−)-epicatechin, ferulic acid, homogentisic acid, polydatin, syringaldehyde, taxifolin, umbelliferon, sinapic acid, kaempferol and vanillic acid, were monitored.

2.4. Fatty Acids Profile

An aliquot (20 mg) of ethanol extract from peel, seed and pulp of each fruit underwent alkaline transmethylation [8]. The gas-chromatography (GC) analysis of fatty acid methyl esters (FAME) was carried out using GC-430 apparatus (Varian, Palo Alto, CA, USA) equipped with a Flame Ionization Detector (FID) and a CPSil88 fused silica capillary column (100 m, 0.25 mm internal diameter, film thickness 0.2 µm, Chrompack, Middelburg, Netherlands). The carrier gas was helium at a flow rate of 1.6 mL/min; the oven temperature program started from 160 °C, rose to 240 °C at a rate of 4 °C/min and then remained at 240 °C for 10 min. The injector temperature was 260 °C. The sample was injected into a split/splitless system. Peaks were identified by comparison with known standards.

2.5. ABTS and DPPH Radical Scavenging Assays

The radical scavenging potential was investigated by using two different spectrophotometric methods: 2,2′-azino-bis (3-ethylbenzothiazoline-6-sulphonic acid) (ABTS) and 2,2-diphenyl-1-picrylhydrazyl (DPPH) assays [9]. The radicals (ABTS or DPPH) scavenging ability was calculated as follows:

$$\text{scavenging activity} = [(A_0 - A)/A_0] \times 100$$

where A_0 is the absorbance of the control reaction and A is the absorbance in the presence of the extract. Ascorbic acid was used as positive control.

2.6. β-Carotene Bleaching Test

The protection of extract on lipid peroxidation was measured as previously described [9]. Briefly, β-carotene solution was added to linoleic acid and 100% Tween 20. The absorbance of the samples, standard and control was measured at 470 nm against a blank at t = 0 and successively at 30 and 60 min. Propyl gallate was used as positive control.

2.7. FRAP (Ferric Reducing Ability Power) Assay

The FRAP assay was applied following the procedure previously described [7]. The FRAP value represents the ratio between the slope of the linear plot for reducing Fe^{3+}-TPTZ reagent by different

Colombian fruits extract compared to the slope of the plot for FeSO$_4$. Butylated hydroxytoluene (BHT) was used as positive control.

2.8. Relative Antioxidant Capacity Index (RACI) Calculation

Relative antioxidant capacity index (RACI) is a statistical application to integrate the antioxidant capacity values generated from different in vitro methods [10]. Thus, data obtained from ABTS, DPPH, β-carotene bleaching tests and FRAP tests were used to calculate RACI value for samples. The standard score is calculated by using the following Equation:

$$(x - \mu)/\sigma$$

where x is the raw data, μ is the mean and σ is the standard deviation.

2.9. Global Antioxidant Score (GAS)

For each sample, the average of T-scores was used to calculate the GAS value. T-score is calculated by the following Equation:
$$\text{T-score} = (X - \min)/(\max - \min)$$

where min and max, respectively, represent the smallest and largest values of variable X among the investigated extract [10].

2.10. α-Amylase and α-Glucosidase Inhibitory Assays

The α-amylase inhibitory test was performed as previously described [11]. Concisely, α-amylase solution, starch solution and colorimetric reagent were prepared. Control and samples at different concentrations were added to starch solution and left to react with the enzyme at room temperature for 5 min. The generation of maltose was quantified at 540 nm by the reduction of 3,5-dinitrosalicylic acid to 3-amino-5-nitrosalicylic acid.

The α-glucosidase inhibition was measured as previously described [11]. Both control and samples (at concentrations in the range 0.01–1 mg/mL) were added to maltose solution and left to equilibrate at 37 °C. The reaction was started by adding the enzyme and left to incubate at 37 °C for 30 min. A perchloric acid solution was used to stop the reaction. The supernatant was collected and mixed with peroxidase/glucose oxidase and o-dianisidine and left to incubate for 30 min at 37 °C. The absorbance was measured at 500 nm. Acarbose was the positive control in both tests.

2.11. Statistical Analysis

The inhibitory concentration 50% (IC_{50}) was calculated by non-linear with the use of Prism Graphpad Prism version 4.0 for Windows, GraphPad Software, San Diego, CA, USA. Differences within and between groups were evaluated by one-way ANOVA followed by a multicomparison Dunnett's test (α = 0.05): **** $p < 0.0001$, *** $p < 0.001$, ** $p < 0.05$, compared with the positive controls.

The concentration-response curve was obtained by plotting the percentage of inhibition versus the concentrations.

3. Results and Discussion

3.1. UHPLC-ESI-HRMS Phenolic Profile

The phenolic profile of the ethanol extracts of Colombian fruits was outlined by UHPLC-ESI-HRMS analysis (Table 1). A total of 16 phenolic compounds were detected. Concerning the Solanaceae family, *C. betacea* showed the highest phenol concentrations. In all fruits, the peel extracts presented higher phenol amounts than those derived from pulp and seed. Despite a different content, chlorogenic acid resulted in the most abundant compound in all extracts. An appreciable amount of rutin hydrate was found in the peel extract from *S. quitoense*. This finding was in agreement with data reported by Gancel et al. [12], who found that the main chlorogenic acid and dihydrocaffeoyl spermidine were located in all parts of *S. quitoense*, whereas the flavonol glycosides were exclusively present in the peel. *trans*-Cinnamic acid was recorded only in the *C. betacea* pulp extract. Moreover, although at low concentrations, ferulic acid and sinapic acid were revealed exclusively in peel and pulp extracts from *C. betacea*. A different phenolic profile was reported in *C. betacea* pulp (yellow variety) by Mertz et al. [13]. Besides chlorogenic acids, they found glycoside ester of caffeic and ferulic acids. These differences could be related to the different solvent used for phenolic extraction. In fact, acetone was used as extraction solvent [13]. Markedly different phenolic composition occurred in extracts from fruits belonging to Passifloraceae family. *P. pinnatistipula* peel extract showed *trans*-cinnamic acid as main component, followed by sinapic and gallic acids. In pulp extract, only two compounds were identified such as polydatin and rutin hydrate. The most abundant compound identified in seed extract was (−)-epicatechin. Differently, the peel extract of *P. tripartita* showed ferulic acid and (+)-catechin as main compounds. (+)-Catechin was more abundant in pulp and seed extracts, respectively. Moreover, the pulp was characterised by a high content of sinapic acid (211.4 ppm). The other identified compounds were in the range 0.2–11.9 ppm. Regarding *P. ligularis* fruit, the peel extract was mainly formed by (+)-catechin, whereas the pulp+seed extract was characterised by low total phenolic amounts (<5 ppm) with polydatin and vanillin as the main compounds. Unfortunately, no data about phenolic profile in the literature were found for the investigated Passifloraceae fruits.

3.2. Fatty Acid Composition

The average fatty acid composition of extracts from peel, pulp and seed of the investigated fruits is reported in Table 2. The fatty acid composition of the extracts changed according to the family fruit and to the part of the fruit. Generally, all extracts were mainly formed by essential fatty acids, such as α-linoleic (C18:2 n-6) and α-linolenic acid (C18:3 n-3) acids. In detail, linoleic acid represented the most abundant fatty acid in all Passifloraceae fruit extracts, except in the pulp extract from the *P. tripartita* fruit. This extract did not contain monounsaturated fatty acids. It presented four fatty acids: palmitic (C16:0), stearic (C18:0), linoleic and linolenic acids. The latter was the most abundant.

Concerning the Solanaceae family, α-linolenic acid (C18:3 n-6) was exclusively found in extracts from *P. peruviana*. Moreover, all extracts from *S. quitoense* and *P. peruviana* presented α-linolenic acid as major fatty acid. Differently, peel and pulp extracts from *C. betacea* presented the oleic acid (C18:1 n-9) as the predominant fatty acid, whereas the major fatty acid in seed extract was α-linoleic acid. In a previously study, Ramakrishnan et al. [14] investigated seed oil of *C. betacea* and they found that the most abundant fatty acids were linoleic (70.47%) and oleic (14.93%) acids.

3.3. Antioxidant Activity

Different tests were applied to screen the antioxidant activity of Colombian fruits. A concentration-effects relationship was found for all tested extracts in all tests except for FRAP assay (Table 3). DPPH and ABTS test were used to screen the radical scavenging potential. Among the Solanaceae family, the *S. quitoense* peel showed the highest DPPH radical scavenging potential with an IC_{50} of 38.8 µg/mL followed by *C. betacea* seed (IC_{50} of 57.9 µg/mL). Higher IC_{50} values were found when an ABTS assay was used. A promising protection of lipid peroxidation was observed with *S. quitoense* pulp+seed (IC_{50} of 6.9 µg/mL) followed by *P. peruviana* peel (IC_{50} of 10.2 µg/mL).

Table 1. Quantification (mg/kg ethanol extract) of identified phenolic compounds in Colombian fruits from Solanaceae and Passifloraceae families.

	(+)-Catechin	Chlorogenic Acid	trans-Cinnamic Acid	p-Coumaric Acid	(−)-Epicatechin	Ferulic Acid	Gallic Acid	Homogentisic Acid	Polydatin	Rutin Hydrate	Sinapic Acid	Syringaldehyde	Taxifolin	Vanillic Acid	Vanillin	Veratric Acid	ΣPhenolic
Solanaceae																	
S. quitoense																	
Peel	n.d.	98.6 ± 2.9	n.d.	0.4 ± 0.0	n.d.	n.d.	1.3 ± 0.0	n.d.	n.d.	51.1 ± 3.6	n.d.	n.d.	0.3 ± 0.0	n.d.	n.d.	n.d.	151.6 ± 2.5
Pulp + Seeds	n.d.	19.1 ± 1.0	n.d.	n.d.	n.d.	n.d.	n.d.	n.d.	n.d.	n.d.	n.d.	n.d.	0.1 ± 0.0	n.d.	n.d.	n.d.	20.0 ± 1.0
P. peruviana																	
Peel	n.d.	1.5 ± 0.1	n.d.	n.d.	n.d.	n.d.	0.9 ± 0.0	n.d.	0.6 ± 0.03	0.1 ± 0.0	n.d.	n.d.	0.2 ± 0.0	n.d.	n.d.	n.d.	3.2 ± 0.2
Pulp + Seeds	n.d.	0.7 ± 0.1	n.d.	n.d.	n.d.	n.d.	n.d.	n.d.	0.6 ± 0.07	n.d.	n.d.	n.d.	0.1 ± 0.0	n.d.	n.d.	n.d.	1.4 ± 0.2
C. betacea																	
Peel	n.d.	253.8 ± 3.8	n.d.	0.2 ± 0.0	n.d.	8.7 ± 0.2	n.d.	n.d.	n.d.	9.5 ± 1.4	10.3 ± 0.7	0.7 ± 0.0	0.2 ± 0.0	n.d.	n.d.	n.d.	284.1 ± 1.2
Pulp	n.d.	125.5 ± 1.2	28.7 ± 2.0	0.2 ± 0.0	n.d.	7.6 ± 0.1	n.d.	n.d.	n.d.	n.d.	1.6 ± 0.2	n.d.	0.1 ± 0.0	n.d.	n.d.	n.d.	165.1 ± 2.2
Seeds	0.8 ± 0.2	37.7 ± 1.3	n.d.	n.d.	2.5 ± 0.1	n.d.	n.d.	n.d.	n.d.	0.7 ± 0.0	n.d.	n.d.	0.1 ± 0.0	n.d.	n.d.	n.d.	42.1 ± 1.2
Passifloraceae																	
P. pinnatistipula																	
Peel	n.d.	n.d.	132.4 ± 5.1	1.5 ± 0.0	2.4 ± 0.2	4.6 ± 0.5	8.1 ± 0.2	n.d.	1.1 ± 0.03	4.7 ± 0.2	9.4 ± 0.1	n.d.	1.2 ± 0.0	n.d.	n.d.	n.d.	166.2 ± 3.2
Pulp	n.d.	n.d.	n.d.	n.d.	n.d.	n.d.	n.d.	n.d.	0.5 ± 0.02	0.4 ± 0.1	n.d.	n.d.	n.d.	n.d.	n.d.	n.d.	1.0 ± 0.1
Seeds	n.d.	n.d.	n.d.	0.3 ± 0.0	8.1 ± 0.8	n.d.	n.d.	n.d.	3.4 ± 0.22	n.d.	n.d.	n.d.	0.2 ± 0.0	n.d.	n.d.	n.d.	12.4 ± 0.3
P. tripartita																	
Peel	40.1 ± 2.0	n.d.	n.d.	2.0 ± 0.1	n.d.	47.8 ± 1.8	n.d.	n.d.	0.7 ± 0.05	n.d.	10.2 ± 0.9	n.d.	n.d.	n.d.	n.d.	n.d.	101.9 ± 1.3
Pulp	140.9 ± 8.2	n.d.	n.d.	1.8 ± 0.2	7.9 ± 0.5	16.3 ± 1.4	n.d.	n.d.	1.9 ± 0.27	n.d.	211.4 ± 1.5	0.8 ± 0.0	n.d.	0.5 ± 0.0	n.d.	n.d.	383.3 ± 1.3
Seeds	208.8 ± 3.6	n.d.	n.d.	n.d.	81.5 ± 3.5	n.d.	n.d.	n.d.	0.7 ± 0.03	n.d.	55.3 ± 4.1	2.2 ± 0.1	2.1 ± 0.1	1.3 ± 0.0	n.d.	42.6 ± 1.7	394.2 ± 2.6
P. ligularis																	
Peel	257.6 ± 4.2	0.2 ± 0.0	n.d.	1.6 ± 0.1	n.d.	11.9 ± 1.9	2.4 ± 0.3	8.95 ± 1.1	2.1 ± 0.08	n.d.	4.0 ± 0.2	2.6 ± 0.1	n.d.	n.d.	5.2 ± 0.1	n.d.	297.5 ± 1.4
Pulp + Seeds	n.d.	n.d.	n.d.	0.7 ± 0.1	n.d.	n.d.	n.d.	n.d.	1.7 ± 0.04	0.3 ± 0.0	n.d.	n.d.	n.d.	n.d.	1.8 ± 0.0	n.d.	4.4 ± 0.9

n.d.: not detectable (<LOD); Limit of detection (LOD): (+)-catechin: 10 µg/kg; chlorogenic acid: 30 µg/kg; trans-cinnamic acid: 20 µg/kg; p-coumaric acid: 18 µg/kg; (−)-epicatechin: 10 µg/kg; ferulic acid: 50 µg/kg; gallic acid: 27 µg/kg; homogentisic acid: 30 µg/kg; polydatin: 10 µg/kg; rutin: 10 µg/kg; sinapic acid: 30 µg/kg; syringic acid: 50 µg/kg; taxifolin: 10 µg/kg; vanillic acid: 35 µg/kg; vanillin: 50 µg/kg; veratric acid: 50 µg/kg.

Table 2. Fatty acid (% of total fatty acids) composition of peel, pulp and seed of Colombian fruits from Solanaceae and Passifloraceae families.

Fatty Acids	*S. Quitoense*		*P. Peruviana*		*C. Betacea*			*P. Pinnatistipula*			*P. Tripartita*			*P. Ligularis*		
	Peel	Pulp + Seed	Peel	Pulp + Seed	Peel	Pulp	Seed	Peel	Pulp	Seed	Peel	Pulp	Seed	Peel	Seed	Pulp + Seed
C16:0	22.6 ± 0.1	25.6 ± 0.4	20.6 ± 2.6	19.5 ± 0.7	22.9 ± 0.4	19.8 ± 0.1	13.6 ± 0.0	15.4 ± 0.1	15.6 ± 0.1	11.8 ± 0.0	24.0 ± 0.2	19.1 ± 0.0	9.6 ± 0.0	34.0 ± 0.2	7.8 ± 0.0
C18:0	6.3 ± 0.2	4.7 ± 0.2	2.7 ± 0.1	2.4 ± 0.1	2.6 ± 0.2	1.4 ± 0.1	3.3 ± 0.1	2.2 ± 0.0	2.9 ± 0.1	1.9 ± 0.0	n.d.	25.0 ± 0.1	1.9 ± 0.0	2.1 ± 0.0	2.5 ± 0.0
C20:0	n.d.	n.d.	5.4 ± 0.5	n.d.	n.d.	n.d.	n.d.	n.d.	n.d.	n.d.	n.d.	n.d.	n.d.	n.d.	n.d.
SFA	28.9 ± 0.1	30.3 ± 0.6	28.8 ± 2.1	22.0 ± 0.5	25.4 ± 0.2	21.5 ± 0.2	16.9 ± 0.1	17.6 ± 0.1	18.5 ± 0.2	13.6 ± 0.0	24.0 ± 0.2	44.1 ± 0.1	11.5 ± 0.0	36.2 ± 0.2	10.3 ± 0.1
C16:1	1.6 ± 0.1	n.d.	0.6 ± 0.0	1.4 ± 0.1	n.d.	1.8 ± 0.1	0.7 ± 0.0	0.4 ± 0.0	4.7 ± 0.1	0.2 ± 0.0	n.d.	n.d.	n.d.	n.d.	0.2 ± 0.0
C18:1 9	12.7 ± 0.3	17.0 ± 0.1	26.5 ± 0.5	20.5 ± 0.5	41.4 ± 0.3	39.1 ± 0.2	17.2 ± 0.6	11.3 ± 0.0	7.5 ± 0.1	13.6 ± 0.0	24.0 ± 0.2	n.d.	10.8 ± 0.1	3.3 ± 0.0	15.1 ± 0.3
C18:1 11	3.6 ± 0.3	n.d.	6.4 ± 0.1	8.4 ± 0.4	2.3 ± 0.1	1.2 ± 0.1	1.4 ± 0.1	1.2 ± 0.0	8.1 ± 0.1	10.0 ± 0.0	5.7 ± 0.3	n.d.	n.d.	0.9 ± 0.1	0.3 ± 0.0
MUFA	18.0 ± 0.7	17.0 ± 0.1	33.5 ± 0.6	30.3 ± 1.0	43.7 ± 0.4	41.7 ± 0.4	19.3 ± 0.8	13.0 ± 0.1	20.3 ± 0.2	10.4 ± 0.3	5.7 ± 0.3	n.d.	10.8 ± 0.1	4.2 ± 0.1	15.6 ± 0.4
C18:2 n-6	11.5 ± 0.1	23.5 ± 0.9	3.6 ± 0.1	11.9 ± 0.1	10.4 ± 0.1	13.1 ± 0.2	58.3 ± 1.0	63.9 ± 0.3	41.0 ± 0.4	75.5 ± 1.4	39.9 ± 0.3	16.6 ± 0.0	76.6 ± 0.3	36.0 ± 0.1	73.3 ± 0.7
C18:3 n-6	n.d.	n.d.	4.3 ± 0.1	4.4 ± 0.7	n.d.	n.d.	n.d.	n.d.	n.d.	n.d.	n.d.	n.d.	n.d.	n.d.	n.d.
C18:3 n-3	41.6 ± 0.9	29.2 ± 0.5	29.8 ± 1.3	31.4 ± 1.0	20.5 ± 0.4	23.6 ± 0.7	5.5 ± 0.1	5.5 ± 0.0	20.2 ± 0.0	0.5 ± 0.0	30.4 ± 0.2	39.3 ± 0.1	1.0 ± 0.0	23.6 ± 0.0	0.8 ± 0.0
PUFA	53.1 ± 0.8	52.7 ± 0.5	37.7 ± 1.5	47.7 ± 0.4	30.9 ± 0.3	36.8 ± 0.5	63.8 ± 0.9	69.4 ± 0.2	61.2 ± 0.4	76.0 ± 0.0	70.4 ± 0.1	55.9 ± 0.1	77.7 ± 0.1	59.6 ± 0.1	74.1 ± 1.4

SFA, saturated fatty acid; MUFA, monounsaturated fatty acid; PUFA, polyunsaturated fatty acid; n.d.; not detectable. Results represents means ± standard deviation (S.D.) ($n = 3$).

Table 3. In vitro antioxidant and hypoglycaemic activity of selected Colombian fruits belonging to Solanaceae and Passifloraceae family.

Title	Sample	DPPH (IC$_{50}$ μg/mL)	ABTS (IC$_{50}$ μg/mL)	β-Carotene Bleaching Test (IC$_{50}$ μg/mL)	FRAP (μM Fe(II)/g)	RACI	GAS	α-Amylase (IC$_{50}$ μg/mL)	α-Glucosidase (IC$_{50}$ μg/mL)
				Solanaceae					
S. quitoense	Peel	38.8 ± 2.1 ****	167.6 ± 3.7 ****	11.1 ± 1.3 ****	49.4 ± 1.5 ****	−0.43	0.81	31.8 ± 1.2 ****	27.9 ± 1.0
	Pulp + Seed	61.3 ± 2.0 ****	576.8 ± 7.5 ****	6.9 ± 0.5	16.2 ± 2.5 ****	−0.38	0.87	54.9 ± 2.4	57.1 ± 2.0 ****
P. peruviana	Peel	117.9 ± 5.1 ****	843.3 ± 3.9 ****	10.2 ± 1.1 ***	13.0 ± 0.8 ****	0.20	1.72	34.1 ± 2.2 ****	37.6 ± 2.5
	Pulp + Seed	65.3 ± 2.1 ****	>1000	19.3 ± 1.6 ****	7.3 ± 0.6 ****	−0.03	1.40	64.3 ± 3.0 ****	45.0 ± 1.1
C. betacea	Peel	74.7 ± 3.5 ****	149.8 ± 3.6 ****	21.9 ± 1.5 ****	63.9 ± 3.5	0.05	1.52	77.1 ± 3.3 ****	32.9 ± 2.9 **
	Pulp	141.3 ± 5.5 ****	463.8 ± 4.7 ****	93.8 ± 4.5 ****	9.8 ± 0.6 ****	0.68	2.42	92.7 ± 3.8 ****	95.1 ± 4.6 ****
	Seed	57.9 ± 1.5 ****	329.8 ± 3.3 ****	58.2 ± 4.0 ****	25.4 ± 2.5 ****	−0.09	1.30	102.9 ± 4.0 ****	195.1 ± 4.7 ****
				Passifloraceae					
P. pinnatistipula	Peel	207.9 ± 2.5 ****	125.3 ± 1.5 ****	>1000	40.5 ± 1.4 ****	0.23	1.47	46.4 ± 2.2	37.7 ± 2.2
	Pulp	671.9 ± 5.7 ****	151.7 ± 1.7 ****	50.6 ± 2.5 ****	3.2 ± 0.7 ****	0.11	1.15	78.2 ± 3.7 ****	44.1 ± 2.3
	Seed	372.2 ± 4.0 ****	>1000	133.8 ± 2.7 ****	28.7 ± 2.3 ****	0.59	1.74	54.1 ± 2.5	40.6 ± 2.8
P. tripartita	Peel	3.9 ± 0.8	177.8 ± 3.2 ****	>1000	22.7 ± 2.2 ****	0.03	1.18	86.3 ± 3.0 ****	56.1 ± 3.1 ****
	Pulp	3.8 ± 0.5	50.1 ± 2.5 ****	3.8 ± 0.3 ****	64.1 ± 3.4	−0.59	0.62	67.5 ± 3.7 ****	78.4 ± 3.9 ****
	Seed	3.2 ± 0.2	96.2 ± 3.7 ****	14.8 ± 0.7 ****	92.0 ± 3.7 ****	−0.49	0.27	52.7 ± 3.5	54.6 ± 3.2 ****
P. ligularis	Peel	61.3 ± 2.2 ****	282.1 ± 4.0 ****	265.1 ± 4.0 ****	39.9 ± 4.3 ****	−0.22	0.69	122.7 ± 4.5 ****	154.9 ± 3.9 ****
	Pulp + Seed	73.9 ± 2.7 ****	223.9 ± 3.7 ****	116.0 ± 3.1 ****	42.9 ± 3.8 ****	0.35	1.40	22.6 ± 3.7 ****	24.8 ± 3.9 **
Positive controls	Propyl gallate [a] Ascorbic acid [a] BHT [a] Acarbose [a]	2.0 ± 0.01	1.7 ± 0.8	1.0 ± 0.01	63.2 ± 4.5			50.0 ± 0.9	35.5 ± 1.2

Data are given as media ± S.D. ($n = 3$); 2,2-diphenyl-1-picrylhydrazyl (DPPH) Radical Scavenging Activity Assay; Antioxidant Capacity Determined by Radical Cation 2,2′-Azino-bis(3-ethylbenzothiazoline-6-sulfonic acid) (ABTS$^{\bullet+}$), β-Carotene bleaching test, Ferric Reducing Ability Power (FRAP); Relative antioxidant capacity Index (RACI); Global Antioxidant Score (GAS); [a] Propyl gallate, ascorbic acid and butylated hydroxytoluene (BHT) were used as positive control in antioxidant test, while acarbose was used in carbohydrate hydrolysing enzyme inhibition assays. Differences within and between groups were evaluated by one-way ANOVA followed by a multicomparison Dunnett's test (=0.05): **** $p < 0.0001$, *** $p < 0.001$, ** $p < 0.05$ compared with the positive controls.

C. betacea peel exhibited a FRAP value similar to those reported for the positive control BHT (63.9 versus 63.2 μM Fe(II)/g). The antioxidant potential of *C. betacea* non-edible portion in terms of radicals scavenging ability and ferric reducing agent was previously described [15]. Several works evidenced also the antioxidant potential of *P. peruviana* whole fruit [9].

Considering Passifloraceae family, *P. tripartita* extracts showed the most promising DPPH radical scavenging activity with IC_{50} values of 3.2, 3.8, and 3.9 μg/mL for seed, pulp and peel, respectively. Additionally, peels exhibited an interesting $ABTS^+$· radical scavenging potential (IC_{50} value of 50.1 μg/mL) and the best protection of lipid peroxidation (IC_{50} value of 3.8 μg/mL). Moreover, *P. tripartita* showed the highest ferric reducing ability and in particular seed exhibited a 1.4-time higher FRAP value than those reported for BHT. Analysis of the chemical profile revealed a high content of sinapic acid, (−)-epicatechin and ferulic acid in the pulp, seed and peel, respectively.

To get a ranking of Colombian fruits antioxidant capacity, relative antioxidant capacity index (RACI) and Global Antioxidant Score (GAS) were calculated. Both statistical applications were generated from the perspective of statistics by integrating the antioxidant capacity values generated from different in vitro methods (Table 3). Based on RACI and GAS value among fruits derived from Solanaceae family, *S. quitoense* peel had the highest antioxidant potential, while *P. tripartita* pulp exhibited the highest antioxidant capacity among investigated Passifloraceae fruits. Pearson's correlations coefficient evidenced a positive correlation between the content of identified phenol and FRAP value for both Solanaceae and Passifloraceae (r = 0.79 and 0.81, respectively). Phenolic compounds showed redox properties acting as reducing agents, hydrogen donors and singlet oxygen quenchers [16]. Selected Colombian fruits have been shown to be a good source of phenols that are known to possess a great antioxidant potential. In particular, these phytochemicals are able to act as free radical scavengers, metal chelator and to protect lipids from peroxidation. Moreover, the antioxidant activity exerted by phenols is the results of influence on cell signalling pathways and on gene expression. The antioxidant activity results from a complex interaction between different compounds in phytocomplex, which produce synergistic effect.

3.4. Carbohydrate Hydrolysing Enzymes Inhibition by Colombian Fruits Extracts

The inhibition of carbohydrates-hydrolysing enzymes α-amylase and α-glucosidase was investigated and results are herein reported. All investigated samples inhibited both enzymes in a concentration-dependent manner, however, generally, α-glucosidase showed to be more sensible (Table 3). An analysis of the data evidenced that *S. quitoense* peel extract had the best activity with IC_{50} values of 27.9 and 31.8 μg/mL for α-glucosidase and α-amylase, respectively. Additionally, a promising hypoglycaemic activity was observed with *P. peruviana* peel that showed IC_{50} values of 34.1 and 37.6 μg/mL against α-amylase and α-glucosidase, respectively. A lower IC_{50} value was found with *C. betacea* peel against α-glucosidase (IC_{50} of 32.9 μg/mL). Among Passiflorareae fruits, *P. ligularis* pulp + seed extract showed the highest activity with IC_{50} values of IC_{50} values of 22.6 and 24.8 μg/mL against α-amylase and α-glucosidase, respectively, followed by *P. pinnatistipula* (IC_{50} values of 46.4 and 37.7 μg/mL against α-amylase and α-glucosidase, respectively). Several samples showed a lower IC_{50} value than those reported for the largely prescribed drugs acarbose herein used as positive control.

Among the identified phytochemicals, flavonoids were mainly involved in the management of Type 2 Diabetes Mellitus (T2DM). These compounds were able to (a) inhibit carbohydrates-hydrolysing enzymes; (b) inhibit sodium-dependent glucose transporter 1 (SGLT1); (c) stimulate insulin secretion; (d) reduce hepatic glucose output; (e) enhance insulin-dependent glucose uptake. In particular, rutin, which was particularly abundant in *S. quitoense* peel extract, inhibited both -amylase and α-glucosidase with IC_{50} values of 0.043 and 0.037 μM, respectively [17]. A similar consideration could be done for chlorogenic acid that inhibited both α-amylase and α-glucosidase with IC_{50} values of 25 and 26.1 μM, respectively [18]. Different hypoglycaemic mechanisms were reported in the literature for this phytochemical. It stimulates glucose uptake in skeletal muscle and suppression of hepatic glucose production by the activation of AMPK. In addition, it has been found that it could modulate glucose in both genetically and healthy metabolic related disorders including DM [19]. Moreover, *S. quitoense* peel extract was rich in linoleic acid. This acid showed a more potent α-glucosidase inhibitory activity than acarbose, whereas a weaker anti α-amylase activity was observed [20].

4. Conclusions

In the present study, we investigated the phenols and fatty acids profile and in vitro antioxidant and hypoglycaemic activity of edible and non-edible parts of Colombian fruits from Solanaceae and Passifloraceae families. The phenolic profile of the ethanol extracts of selected exotic fruits was reported herein for the first time. A total of 16 phenolic compounds were detected, with peel extracts showing the highest amount of phenolics compared to pulp and seed samples. Generally, chlorogenic acid and rutin hydrate were the most representative compounds. Considering only the fruit species in which it was possible to separate peel, pulp and seed, the results showed how the omega-3 fatty acids preferentially accumulated in the pulp and in the peel rather than in the seeds. Furthermore, it was found that *S. quitoense*, which contains considerable amount of rutin hydrate, had the highest antioxidant potential among the Solanaceae samples, while *P. tripartita* fruits exhibited the highest antioxidant capacity among the investigated Passifloraceae samples. Moreover, carbohydrate-hydrolysing enzyme inhibitory activity studies highlighted a promising hypoglycaemic activity of *P. ligularis* seed and *S. quitoense* peel samples. The latter demonstrated that carbohydrates-hydrolysing enzymes inhibition was higher than the commercial drug acarbose. Although some identified flavonoids possess hypoglycaemic activity, it is not possible to attribute these interesting properties to these compounds, but rather to the phytocomplex in which a synergism of action can occur.

Overall, the present results supported the possibility of using peels or seeds, which are normally considered as food by-products of juice production, as potential sources of bioactive ingredients in functional beverages, nutraceutical or dietary supplement formulations for the treatment and/or prevention of several diseases associated with oxidative stress such as type 2 diabetes.

Author Contributions: Conceptualization, P.L.; Data curation, N.G.F.; Formal analysis, R.T.; Investigation, M.R.L., O.N., R.T., M.B., D.F. and D.P.; Project administration, P.L.; Supervision, M.R.L., P.L., D.P. and O.N.; Visualization, M.B. and D.P.; Writing—original draft, M.R.L. and D.P.; Writing—review & editing, L.C., S.M. and E.M.

Funding: This research received no external funding

Acknowledgments: The authors are grateful to the Spanish Ministry of Economy and Competitiveness under the project CTQ2015-63968-C2-1-P. Daria Filatova is grateful to the European Commission for the category A Fellowship for Erasmus Mundus master program.

Conflicts of Interest: The authors declare that they have no conflict of interest.

References

1. International Diabetes Federation-Home. Available online: https://www.idf.org/ (accessed on 23 January 2019).
2. Montgomery, M.K.; Turner, N. Mitochondrial dysfunction and insulin resistance: An update. *Endocr. Connect.* **2015**, *4*, R1–R15. [CrossRef] [PubMed]
3. Loizzo, M.R.; Bonesi, M.; Nabavi, S.M.; Sobarzo-Sánchez, E.; Rastrelli, L.; Tundis, R. Hypoglycaemic Effects of plants food constituents *via* inhibition of carbohydrate-hydrolysing enzymes: From chemistry to future applications. *Nat. Prod. Target. Clin. Relev. Enzym.* **2017**, *1*, 135–161.
4. Loizzo, M.R.; Bonesi, M.; Menichini, F.; Tenuta, M.C.; Leporini, M.; Tundis, R. Antioxidant and Carbohydrate-Hydrolysing Enzymes Potential of *Sechium edule* (Jacq.) Swartz (Cucurbitaceae) Peel, Leaves and Pulp Fresh and Processed. *Plant Foods Hum. Nutr.* **2016**, *71*, 381–387. [CrossRef] [PubMed]
5. Tundis, R.; Bonesi, M.; Sicari, V.; Pellicanò, T.M.; Tenuta, M.C.; Leporini, M.; Menichini, F.; Loizzo, M.R. *Poncirus trifoliata* (L.) Raf.: Chemical composition, antioxidant properties and hypoglycaemic activity *via* the inhibition of alpha-amylase and alpha-glucosidase enzymes. *J. Funct. Foods* **2016**, *25*, 477–485. [CrossRef]
6. Heiser Charles, B., Jr. Some Ecuadorian and Colombian Solanums with edible fruits. [Algunos Solanums Colombianos y Ecuatorianos con frutos comestibles. *Cienc. Nat.* **1968**, *11*, 3–9.

7. Facciola, S. *Cornucopia II: A Source Book of Edible Plants*; Kampong Publications: Vista, CA, USA, 1998.
8. Suter, B.; Grob, K.; Pacciarelli, B. Determination of fat content and fatty acid composition through 1-min transesterification in the food sample; principles. *Z. Für Leb. Forsch. A* **1997**, *204*, 252–258. [CrossRef]
9. Loizzo, M.R.; Pacetti, D.; Lucci, P.; Núñez, O.; Menichini, F.; Frega, N.G.; Tundis, R. *Prunus persica* var. *platycarpa* (Tabacchiera Peach): bioactive compounds and antioxidant activity of pulp, peel and seed ethanolic extracts. *Plant Foods Hum. Nutr.* **2015**, *70*, 331–337. [CrossRef] [PubMed]
10. Todorovic, V.; Milenkovic, M.; Vidovic, B.; Todorovic, Z.; Sobajic, S. Correlation between antimicrobial, antioxidant activity, and polyphenols of alkalized/nonalkalized cocoa powders. *J. Food Sci.* **2017**, *82*, 1020–1027. [CrossRef] [PubMed]
11. Loizzo, M.R.; Pugliese, A.; Bonesi, M.; Tenuta, M.C.; Menichini, F.; Xiao, J.; Tundis, R. Edible flowers: A rich source of phytochemicals with antioxidant and hypoglycemic properties. *J. Agric. Food Chem.* **2016**, *64*, 2467–2474. [CrossRef] [PubMed]
12. Gancel, A.-L.; Alter, P.; Dhuique-Mayer, C.; Ruales, J.; Vaillant, F. Identifying carotenoids and phenolic compounds in naranjilla (*Solanum quitoense* Lam. var. Puyo hybrid), an Andean fruit. *J. Agric. Food Chem.* **2008**, *56*, 11890–11899. [CrossRef] [PubMed]
13. Mertz, C.; Gancel, A.-L.; Gunata, Z.; Alter, P.; Dhuique-Mayer, C.; Vaillant, F.; Perez, A.M.; Ruales, J.; Brat, P. Phenolic compounds, carotenoids and antioxidant activity of three tropical fruits. *J. Food Compos. Anal.* **2009**, *22*, 381–387. [CrossRef]
14. Ramakrishnan, Y.; Khoddami, A.; Gannasin, S.P.; Muhammad, K. Tamarillo (*Cyphomandra betacea*) seed oils as a potential source of essential fatty acid for food, cosmetic and pharmacuetical industries. *Acta Hortic.* **2013**, *1012*, 1415–1421. [CrossRef]
15. Hassan, A.; Hawa, S.; Bakar, A.; Fadzelly, M. Antioxidative and anticholinesterase activity of *Cyphomandra betacea* fruit. *Sci. World J.* **2013**, *2013*, 1–7. [CrossRef] [PubMed]
16. Hong, Y.; Lin, S.; Jiang, Y.; Ashraf, M. Variation in contents of total phenolics and flavonoids and antioxidant activities in the leaves of 11 *Eriobotrya* species. *Plant Foods Hum. Nutr.* **2008**, *63*, 200. [CrossRef] [PubMed]
17. Oboh, G.; Ademosun, A.O.; Ayeni, P.O.; Omojokun, O.S.; Bello, F. Comparative effect of quercetin and rutin on α-amylase, α-glucosidase, and some pro-oxidant-induced lipid peroxidation in rat pancreas. *Comp. Clin. Pathol.* **2015**, *24*, 1103–1110. [CrossRef]
18. Oboh, G.; Agunloye, O.M.; Adefegha, S.A.; Akinyemi, A.J.; Ademiluyi, A.O. Caffeic and chlorogenic acids inhibit key enzymes linked to type 2 diabetes (*in vitro*): A comparative study. *J. Basic Clin. Physiol. Pharmacol.* **2015**, *26*, 165–170. [CrossRef] [PubMed]
19. Naveed, M.; Hejazi, V.; Abbas, M.; Kamboh, A.A.; Khan, G.J.; Shumzaid, M.; Ahmad, F.; Babazadeh, D.; FangFang, X.; Modarresi-Ghazani, F. Chlorogenic acid (CGA): A pharmacological review and call for further research. *Biomed. Pharmacother.* **2018**, *97*, 67–74. [CrossRef] [PubMed]
20. Su, C.-H.; Hsu, C.-H.; Ng, L.-T. Inhibitory potential of fatty acids on key enzymes related to type 2 diabetes. *Biofactors* **2013**, *39*, 415–421. [CrossRef] [PubMed]

© 2019 by the authors. Licensee MDPI, Basel, Switzerland. This article is an open access article distributed under the terms and conditions of the Creative Commons Attribution (CC BY) license (http://creativecommons.org/licenses/by/4.0/).

Article

Effect of Artificial LED Light and Far Infrared Irradiation on Phenolic Compound, Isoflavones and Antioxidant Capacity in Soybean (*Glycine max* L.) Sprout

Md Obyedul Kalam Azad [†], Won Woo Kim [†], Cheol Ho Park and Dong Ha Cho *

College of Biomedical Science, Kangwon National University, Chuncheon 24341, Korea; azadokalam@gmail.com (M.O.K.A.); wwkim114@gmail.com (W.W.K.); chpark@kangwon.ac.kr (C.H.P.)
* Correspondence: chodh@kangwon.ac.kr; Tel.: +82-33-250-6475
† Both authors contributed equally.

Received: 1 October 2018; Accepted: 19 October 2018; Published: 22 October 2018

Abstract: The effect of light emitting diode (LED) light and far infrared irradiation (FIR) on total phenol, isoflavones and antioxidant activity were investigated in soybean (*Glycine max* L.) sprout. Artificial blue (470 nm), green (530 nm) LED and florescent light (control) were applied on soybean sprout, from three to seven days after sowing (DAS) in growth chamber. The photosynthetic photon flux density (PPFD) and photoperiod was 150 ± 5 μmol m^{-2}s^{-1} and 16 h, respectively. The FIR was applied for 30, 60 and 120 min at 90, 110 and 130 °C on harvested sprout. Total phenolic content (TP) (59.81 mg/g), antioxidant capacity (AA: 75%, Ferric Reduction Antioxidant Power (FRAP): 1357 μM Fe^{2+}) and total isoflavones content (TIC) (51.1 mg/g) were higher in blue LED compared to control (38.02 mg/g, 58%, 632 μM Fe^{2+} and 30.24 mg/g, respectively). On the other hand, TP (64.23 mg/g), AA (87%), FRAP (1568 μM Fe^{2+}) and TIC (58.98 mg/g) were significantly increased by FIR at 110 °C for 120 min among the treatments. Result suggests that blue LED is the most suitable light to steady accumulation of secondary metabolites (SM) in growing soybean sprout. On the other hand, FIR at 110 °C for 120 min is the best ailment to induce SM in proceed soybean sprout.

Keywords: controlled environment; far infrared irradiation (FIR); light emitting diode (LED) light; flavonoid; soybean sprouts

1. Introduction

Soybean sprouts are one of the most favorable healthy food by consumers in many countries—especially Korea, China and Japan. Sprouting soybean (*Glycine max* L.) are high in quality protein and dietary fiber and containing a lot of functional materials, including isoflavones. Consumption of isoflavones are associated with human health benefits, such as decreased risk of heart disease, menopausal symptoms, cardiovascular disease, as well as breast prostate cancers [1–3]. Isoflavones are categorized chemically according to their functional group, such as glycosides (daidzin, glycitin, genistin) and agylcones (daidzein, glycitein, genistein) (Figure 1).

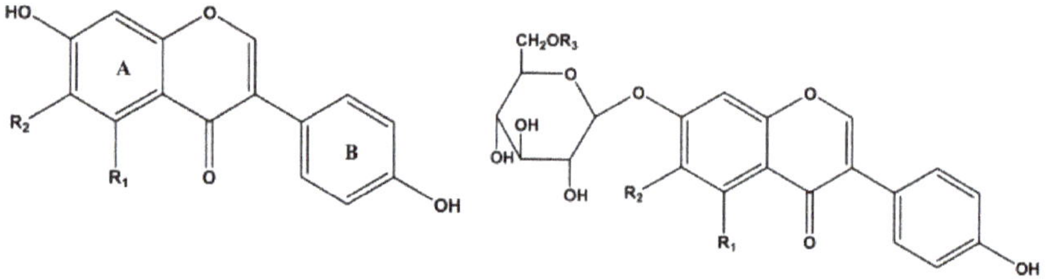

R_1	R_2	Aglycones	R_1	R_2	R_3	Glycosides
H	H	Daidzein	H	H	H	Daidzin
OH	H	Genistein	OH	H	H	Genistin
H	OCH$_3$	Glycitein	H	OCH$_3$	H	Glycitin

Figure 1. Chemical structures of isoflavones.

Artificial light emitting diode (LED) light are being extensively used in controlled production system in order to improve the plant food quality. Light quality directly influences plant growth and chemical composition; therefore, it can be used as an external stimuli to obtain vegetal material with tailored composition [4]. The effects of LED illumination in sprout cultivation has been investigated in several species, such as *Brassica* spp. [5], pea, broccoli, mustard, borage, amaranth, kale, beet, parsley [6], and buckwheat [7].

Artificial blue LED light enhance secondary metabolites, such as ascorbate, total phenolic, anthocyanin, flavonoid contents, and antioxidant activity in basil [6]. Blue light is more efficiently absorbed by photosynthetic pigments than other spectral regions. Sun et al. [7] found that blue light drive CO_2 fixation primarily in the upper palisade mesophyll while green light penetrates deeper and drives CO_2 fixation in the lower palisade and upper spongy mesophyll. Green light is not directly involved in photosynthesis—however, it may affect plant growth and the synthesis of endogenous substances [8]. Swatz et al. [9] suggested that the effects of green light on plant growth and development are similar to those of blue light. Similar positive effects of blue and green light on plant growth, such as photosynthetic capacity and phytochemical production, have been reported on various plants [10,11].

Several reports have shown that the far infrared (FIR) enhanced nutritional quality of the plant foods viz. Chinese herbs, peanut, citrus cakes [12]. Plant secondary metabolites are present as a strong intermolecular covalently bound form with long chain of polymer [13]. The high penetration power of FIR helps the exudation of chemical components in the plant cells and thereby altering biological activity [14]. It is well documented that FIR liberate and activate low-molecular-weighted natural antioxidants in plants [15,16]. Previous researchers studied that FIR significantly increased free radical scavenging capacity in citrus press-cakes and total phenolic content in buckwheat sprouts [17,18].

However, there is not many data available on artificial LED light and FIR effect on phenolic content and antioxidant capacity of soybean sprouts either growing or processing stage, respectively. Therefore, the objectives of this study was to evaluate the effect of LED light and FIR on the accumulation and induce of bioactive compounds in soybean sprouts during growing and processing, respectively.

2. Materials and Methods

2.1. Plant Growth Conditions and LED Light Application

Soybean (*Glycin max* L. var. Seoritae) seeds were ringed with cold water and soaked for 24 h. Then the cleaned seeds were put into a planter having small holes in the bottom. The planter was put in the dark growth chamber. Artificial blue (450–495 nm), green (510–550 nm) LED (Green Power LED Production module. Philips, Poland) and florescent lamps (as a control) were turned on to soybean sprout from two DAS till eight DAS. The photosynthetic photon flux density (PPFD), day/night temperature, relative humidity, CO_2 and a photoperiod of the growth chamber were maintained at 150 ± 5 µmol m^{-2}s^{-1}, 20 ± 0.5 °C, $75 \pm 5\%$, 160 ± 10 ppm and 16 h, respectively. The samples were harvested at 3th, 4th, 5th, 6th and 7th DAS.

2.2. Extraction of Soybean Sprout Grown under LED Lights

The harvested samples were dried using oven at 50 °C and prepared powder using a grinder. Soybean dried powder 1 g were suspended in 100 mL of 80% ethanol and kept over-night in a shaker at room temperature. The extracts were filtered through Advantech 5B filter paper (Tokyo Roshi Kaisha Ltd., Saitama, Japan) and dried using a vacuum rotatory evaporator (EYLA N-1000, Tokyo, Japan) in 40 °C water bath to get crude extract. The crude extract was freeze dried to get moisture content 8–10%. Dried crude extract was diluted using 80% ethanol to prepare 1000 mg/L stock solution and kept at 4 °C for further analysis.

2.3. Application of FIR Irradiation and Extraction Method of Soybean Sprout

An independent experiment was conducted for the application of FIR irradiation on the soybean sprout. In this study, commercially available florescent lamp was used during sprout growing stage. Other climatic factors, photoperiod and duration of the light was same as described above.

Seven days old soybean sprouts were harvested and dried in oven at 50 °C and prepared powder using a grinder. Two grams of powder were mixed with 4 mL of water in a glass petri dish and exposed to FIR dryer (HKD-10; Korea Energy Technology, Seoul, Korea) at 90 °C, 110 °C, 130 °C temperatures, for 30 min, 60 min, 90 min. The treated samples were freeze dried to obtain moisture content 8–10%.

A control sample was collected without FIR application. The FIR treated dried powder and control samples of 1 g were suspended in 100 mL of 80% ethanol and kept over-night in a shaker at room temperature. The extract was filtered and produced crude extract by rotary evaporator as described above. Freeze dried crude extract was diluted using 80% ethanol to prepare 1000 mg/L stock solution.

2.4. Estimation of Total Phenolic Content

Total phenolic content (TP) was determined according to Folin-Ciocalteu assay [19]. In brief, a sample aliquot of 1 mL of stock solution was added to a test tube containing 0.2 mL of phenol reagent (1 M). The volume was increased by adding 1.8 mL of deionized water and the solution was vortexed and left for 3 min for reaction. Furthermore, 0.4 mL of Na_2CO_3 (10% in water, v/v) was added and the final volume (4 mL) was adjusted by adding 0.6 mL of deionized water. The absorbance was measured at 725 nm by spectrophotometer after incubation for 1 h at room temperature. The TP was calculated from a calibration curve using gallic acid standard and expressed as mg of gallic acid equivalent (GAE) per g dry weight basis (dwb).

2.5. Estimation of Isoflavones Content

Total six isoflavones, such as daidzin, glycitin, genistin, daidzein, glycitein and genistein, were analyzed. For the isoflvones estimation, 5 mL of stock solution were centrifuged at 3500 rpm for 10 min. The supernatant was filtered through 0.2 µm pore size syringe filter, type GV (Millipore, Bedford, MA, USA) prior to high performance liquid chromatography injection (HPLC) (Agilent, Stevens Creek Blvd Santa Clara, Santa Clara, CA, USA). Chromatographic separation was performed using a hypersil GOLD RP-18 column (250 mm × 4.00 mm, 5 µm) equipped with a photo diode array detector (Dionex Ultimate PDA-3000, (Dionex Softron GmbH, Dornierstrasse, Germany) at a flow rate of 1 mL min^{-1} at 30 °C with an injection volume of 30 µL. Eluent A contained methanol: Acetonitrile (95:5, v/v) and eluent B contained water-acetic acid (94:6 mL/mL). Linear gradient was used starting with 5% B in A to reach 40% B in A in 25 min. Standard curve and retention times were calibrated using pure standards of soybean isoflavones (Sigma Aldrich Co., St. Louis, MO, USA). All samples were analyzed in triplicate and results were expressed as milligrams per gram (mg/g).

2.6. DPPH Free Radical Scavenging Capacity

The antioxidant capacity was determined on the basis of the scavenging capacity of the stable 2, 2-diphenyl-1 picryl hydrazyl (DPPH) free radical according to methods described by Braca et al. [20] with slight modifications. One mL of stock solution was added to 3 mL of DPPH. The mixture was shaken vigorously and left to stand at room temperature in the dark for 30 min. The absorbance was measured at 517 nm using a spectrophotometer (UV-1800 240 V, Shimadzu Corporation, Kyoto, Japan). The percent inhibition activity were calculated against a blank sample using the following equation: inhibition (%) = (blank sample-extract sample/blank sample) × 100.

2.7. Ferric Reduction Antioxidative Power (FRAP Assay)

The FRAP was determined according to methods described by Benzie and Strain [21]. The FRAP reagent contained 20 mL of a 10 mM TPTZ (2,4,6-tripyridyl-s-triazine) solution in 40 mM HCl, 20 mL of 20 mM iron (III) chloride hexahydrate and 200 mL of 0.3 M acetate buffer at pH 3.6. The FRAP reagent was incubated at 37 °C in a water bath. Aliquots from stock solution of 5 mL were mixed with 45 µL of FRAP reagent. The absorbance was measured at 593 nm using spectrophotometer. Ferrous sulfate heptahydrate was used as a standard for the calibration curve and the results were expressed as FRAP value (µM Fe^{2+}).

2.8. Statistical Analysis

Data are reported as mean ± standard deviation from triplicate analysis. Analysis of variance (ANOVA) accompanied with LSD and Tukey tests (SPSS, Version 15, IBM, New York, NY, USA)) were conducted to identify the significant differences among the samples ($p < 0.05$).

3. Results and Discussion

3.1. Effect of Artificial LED Light on the Accumulation of Total Phenol, Isoflavones Content and Antioxidant Capacity

Light is the important environmental que to improve the bioactive compounds in plant materials [22]. Light stimulate the enzyme activation and regulate the enzyme synthesize pathways, such as PAL (phenylalanine ammonia-lyase) activity in phenyl-propanoid pathways, which promote the bioactive compound accumulation in plant [23]. The isoflvones and total phenol content of the soybean sprouts grown under artificial LED light were shown in Table 1 and Figure 2. It is clearly depicted that isoflavones and total phenol were accumulated higher in the soybean sprout grown under blue LED light compared to the green and florescent light. The isoflavones, such as daidzin, glycitin, genistin, daidzein, glycerin and genistein, were significantly ($p < 0.05$) increased at the five and six DAS. A reduction of the isoflavones contents were observed with increasing DAS. Studies showed that blue light is the most effective lighting source to synthesis flavonoid compounds by stimulating PAL, CHS (chalcone synthesis) and DFR (dihydroflavonol-4-reductage) gene expression [24]. Chi et al. [25], showed that soybean sprouts grown under artificial LED light condition contained higher isoflavones content than those grown under florescent light. Cevallos and Cisneros [26] found higher phenolic content of soybean sprout at seven days of germination. Additionally, Troszyńska et al. [27] reported the highest concentration of polyphenols in lentils at seven days of germination. Conversely, phenolic content in mung bean decreased with germination days [28].

Table 1. Isoflavones contents in soybean sprout grown under different light treatments. (mg/g, dwb).

Treatments	Days after Sowing (DAS)	Daidzin	Glycitin	Genistin	Daidzein	Glycitein	Genistein	Total
Florescent Light (Control)	3	2.74 ± 0.41 e	1.22 ± 0.20 d	4.19 ± 1.21 d	0.22 ± 0.10 d	0.14 ± 0.02 c	0.09 ± 0.01 c	18.32 ± 1.92 c
	4	4.57 ± 0.52 d	1.29 ± 0.28 c	5.44 ± 0.79 d	0.12 ± 0.09 d	0.12 ± 0.05 c	0.11 ± 0.02 c	23.38 ± 1.69 c
	5	4.55 ± 1.22 d	1.09 ± 0.37 d	7.55 ± 1.07 c	0.34 ± 0.25 c	0.21 ± 0.07 a	1.14 ± 0.05 a	28.65 ± 2.94 c
	6	5.94 ± 0.89 d	1.38 ± 0.10 c	6.00 ± 1.32 cd	0.39 ± 0.21 c	0.16 ± 0.09 b	0.96 ± 0.08 b	27.55 ± 2.58 c
	7	4.61 ± 1.08 d	1.52 ± 0.40 c	6.46 ± 0.88 c	0.32 ± 0.21 c	0.18 ± 0.11 ab	0.71 ± 0.11 b	30.24 ± 2.51 b
Blue Light (450–495 nm)	3	8.62 ± 1.20 b	2.09 ± 0.20 a	11.31 ± 1.09 a	0.40 ± 0.04 c	0.17 ± 0.01 ab	1.18 ± 0.01 a	45.88 ± 2.55 a
	4	10.19 ± 1.44 a	1.99 ± 0.08 b	12.61 ± 1.27 a	0.66 ± 0.03 a	0.21 ± 0.01 a	1.29 ± 0.01 a	45.94 ± 2.84 a
	5	9.99 ± 1.07 a	2.15 ± 0.06 a	11.57 ± 1.60 a	0.71 ± 0.02 a	0.10 ± 0.01 c	1.28 ± 0.02 a	51.1 ± 2.76 a
	6	8.52 ± 0.81 b	2.86 ± 0.04 a	12.07 ± 0.85 a	0.77 ± 0.01 a	0.38 ± 0.01 a	1.20 ± 0.04 a	49.12 ± 1.76 a
	7	8.82 ± 1.20 b	2.29 ± 0.20 a	11.31 ± 1.09 a	0.50 ± 0.04 b	0.17 ± 0.01 ab	1.18 ± 0.01 a	47.12 ± 2.55 a
Green Light (510–550 nm)	3	7.16 ± 1.01 c	1.77 ± 0.08 b	9.50 ± 1.08 b	0.31 ± 0.03 c	0.16 ± 0.02 b	1.12 ± 0.01 a	37.38 ± 2.23 b
	4	7.87 ± 0.75 bc	1.87 ± 0.20 b	8.62 ± 0.73 b	0.55 ± 0.04 b	0.12 ± 0.01 c	1.08 ± 0.01 a	37.37 ± 1.74 b
	5	7.25 ± 1.37 c	1.75 ± 0.06 bc	9.52 ± 0.45 b	0.34 ± 0.02 c	0.31 ± 0.01 a	1.27 ± 0.02 a	38.62 ± 1.93 b
	6	7.50 ± 1.06 c	1.87 ± 0.05 b	9.48 ± 1.26 b	0.54 ± 0.01 b	0.09 ± 0.01 d	0.96 ± 0.01 b	38.46 ± 2.40 b
	7	7.18 ± 0.85 c	1.66 ± 0.03 c	7.68 ± 1.04 c	0.44 ± 0.01 b	0.08 ± 0.01 d	0.93 ± 0.01 b	33.58 ± 1.95 b

The values are mean ± standard deviation (SD). ($n = 3$). Values labeled with different letters are significantly different ($p < 0.05$).

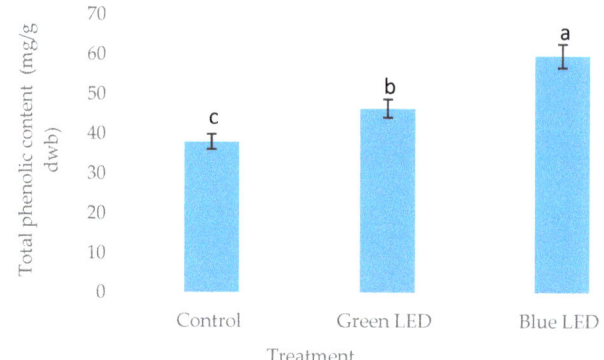

Figure 2. Total phenolic content in soybean sprouts grown under different light. Different lowercase letters within the row indicates significant differences ($p < 0.05$) according to ANOVA. LED: light emitting diode.

The DPPH inhibition activity provide the overall dietary antioxidant ability of the soybean sprout. This measurement is based on the reducing ability of antioxidants towards DPPH radical.

From the Figure 3 it is clearly shown that higher free radical activity (DPPH and FRAP) is demonstrated in blue LED light (75%) compared to green (69%) and control (58%). The FRAP assay showed a large difference in antioxidant reduction profile of soybean sprout in different LED treatments. The FRAP capacity is more than 2 times higher in soybean sprout when grown under blue LED light compared to control. Antioxidant properties of the sprouted vegetables were greatly enhanced by the blue light treatment [29]. Wu et al. [30] noticed that blue light have positive significant effect to improve antioxidant capacity of the sprouted pea seed.

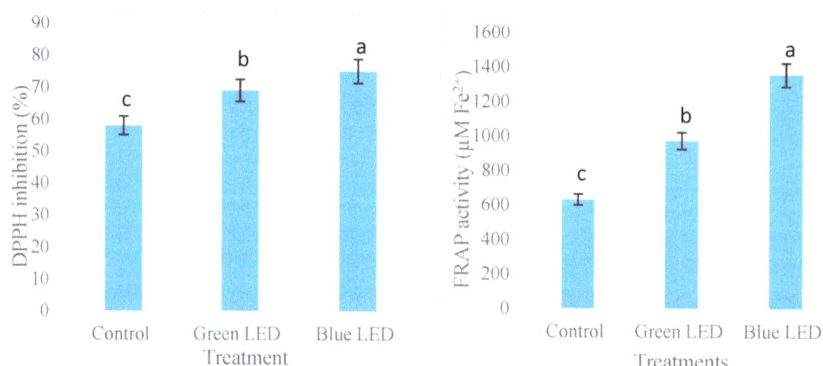

Figure 3. Antioxidant capacity (DPPH and FRAP) of soybean sprout under different light treatment. The values are mean ± SE. ($n = 3$). DPPH, 2-diphenyl-1 picryl hydrazyl; FRAP, Ferric Reduction Antioxidant Power; SE, standard deviation. Different lowercase letters within the row indicates significant differences ($p < 0.05$) according to ANOVA.

The variation of flavonoids during sprouting can be due to catalytic activity of enzymes, such as hydrolases and polyphenoloxidases, which are reported to be activated in sprouting process [29]. Additionally, during the sprouting process, enzyme synthesis occur, which can result in the enhancement of intrinsic secondary compounds.

3.2. Effect of FIR on the Accumulation of Total Phenol, Isoflavones Content and Antioxidant Capacity

Total phenol and isoflavones content of FIR treated soybean sprouts are presented in Table 2 and Figure 3. It is clearly observed that FIR temperature and treatment duration has tremendous effect on TP and isoflavones accumulation in soybean sprout. All the isoflavones, such as daidzinm glycitin, genistin, daidzein, glycitein and genistein, were significantly increased ($p < 0.05$) at 110 °C among the FIR temperature. A significant increase of total isoflavones (58.98 mg/g) were observed in FIR of 110 °C at 120 min, which is nearly 2.3 times higher than the control (25.64 mg/g).

In this study, an application of FIR as a thermal treatment on soybean sprout caused an increase in isoflavones content at 110 °C with exposure time 120 min. Further increase in the FIR temperature decreased the isoflavones content. Increased isoflavones content is due to the transformation of high molecular compounds to low molecular compounds resulting from the breakage of covalent bonds of polymerized polyphenols [12]. A similar trend of increase in phenolic compound, due to different thermal treatment was also observed in different types of fruit and vegetables like sweet corn, ginseng, garlic, tomato, grapes, and onion [14]. The optimum temperature and treatment time of the FIR is depending on the plant materials. Ghimeray et al. [18] observed a highest quantity of the total phenol, total flavonoid and quercetin content of the buckwheat at 120 °C for 60 min of the FIR treatment. It is shown that, the isoflavone content reduced by FIR when exposed for a long time at a high temperature. Similar results were also reported by previous researcher [18].

It is observed in Figure 4 that FIR 110 °C/120 min treatment had the highest FRAP value (1568 µM Fe^{2+}) among the FIR treatments. Comparing our results with the previous results by Ghimeary et al. [18] demonstrated that FIR treatment enhanced DPPH inhibition of the buckwheat. This result is also in agreement with the finding of Randhir er al. [26] where they showed that thermal treatment increased DPPH activity in buckwheat sprout. Our findings are concordant with previous results of total phenol content (Figure 2), showing the phenolics that reacts in Folin–Ciocalteu reaction has a similar interaction in FRAP reaction. Previous study showed higher phenolic and flavonoid content in buckwheat, which was directly correlated with their higher reduction powder [18].

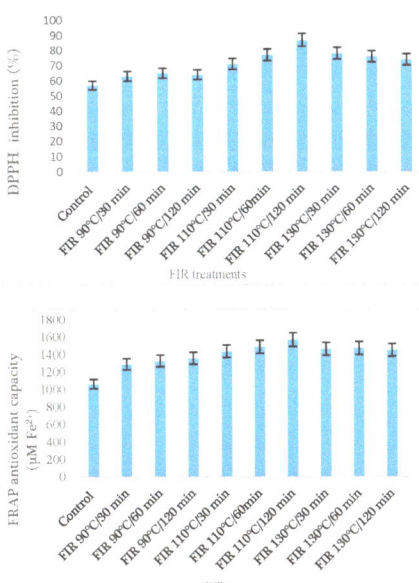

Figure 4. Antioxidant capacity (DPPH and FRAP) of soybean sprout under FIR treatment. The values are mean ± SE. (n = 3). FIR: far infrared irradiation.

Table 2. Isoflavones contents in soybean sprout treated by far infrared irradiation (FIR) treatments. (mg/g, dwb).

Treatment	Daidzin	Glycitin	Genistin	Daidzein	Glycitein	Genistein	Total Isoflavones
Control	4.74 ± 0.41 c	1.22 ± 0.32 c	6.19 ± 1.21 c	0.32 ± 0.10 c	0.25 ± 0.03 cd	0.25 ± 0.01 c	25.69 ± 2.05
FIR-90 °C 30 min	5.19 ± 0.55 c	1.10 ± 0.26 c	5.28 ± 0.13 d	0.57 ± 0.08 b	0.30 ± 0.06 c	1.12 ± 0.11 b	27.12 ± 1.19 c
FIR-90 °C 60 min	5.52 ± 0.29 c	1.26 ± 0.05 c	6.05 ± 0.42 c	0.51 ± 0.10 b	0.29 ± 0.07 c	0.97 ± 0.14 bc	29.2 ± 1.06 c
FIR-90 °C 120 min	5.53 ± 0.30 c	1.19 ± 0.28 c	6.07 ± 0.21 c	0.57 ± 0.03 b	0.37 ± 0.04 c	0.94 ± 0.08 bc	29.34 ± 0.94 c
FIR-110 °C 30 min	7.19 ± 0.27 b	1.49 ± 0.33 c	6.28 ± 0.20 c	0.93 ± 0.26 a	0.73 ± 0.10 b	1.78 ± 0.24 bb	36.8 ± 1.40 b
FIR-110 °C 60 min	8.12 ± 0.29 ab	1.63 ± 0.18 b	6.99 ± 0.46 c	1.29 ± 0.07 a	1.05 ± 0.05 a	2.26 ± 0.32 a	42.68 ± 1.37 b
FIR-110 °C 120 min	10.89 ± 0.24 a	2.39 ± 0.30 a	12.97 ± 0.28 a	0.88 ± 0.13 a	0.94 ± 0.21 a	2.29 ± 0.08 a	58.98 ± 1.24 a
FIR-130 °C 30 min	7.75 ± 0.17 b	1.92 ± 0.19 b	8.72 ± 0.26 b	0.50 ± 0.11 b	0.20 ± 0.03 c	1.42 ± 0.16 b	41.02 ± 0.92 b
FIR-130 °C 60 min	7.15 ± 0.13 b	1.66 ± 0.05 b	8.37 ± 0.14 b	0.32 ± 0.06 c	0.12 ± 0.02 cd	1.25 ± 0.07 b	37.74 ± 0.47 b
FIR-130 °C 120 min	6.88 ± 0.13 b	1.51 ± 0.20 c	6.81 ± 0.37 c	0.24 ± 0.03 c	0.08 ± 0.02 d	1.18 ± 0.09 b	33.4 ± 0.84 c

The values are mean ± SD. ($n = 3$). Values labeled with different letters are significantly different ($p < 0.05$).

4. Conclusions

Growing soybean sprout under artificial blue LED light has shown the highest total phenol, total isoflavones content and antioxidant capacity (DPPH, FRAP). Results also highlighted that FIR thermal treatment increased TP and TIC content in a temperature and exposure time dependent manner. The highest TP and TIC were achieved at 110 °C with exposure time 120 min of FIR. Likewise, scavenging capacity on DPPH and FRAP also increased. The FIR irradiation enhanced aglycone production in soybean sprout, due to the breakage of glycoside bonds of isoflavones in a temperature dependent manner. The greatest increase in bioactive compounds of soybean sprouts was achieved at five to six DAS. The bioactive compounds of soybean sprouts may be optimized by tuning the artificial LED light condition and FIR application strategy during growing and processing, respectively.

Author Contributions: M.O.K.A., W.W.K. performed the experiment, M.O.K.A. wrote the article, C.H.P. revised and improved the manuscript and D.H.C. designed and supervised of the study.

Funding: This research received no external funding.

Acknowledgments: This study was supported by 2014 research grant (No. C1010686-01-01) and Institute of Bioscience and Biotechnology from Kangwon National University, Korea.

Conflicts of Interest: The authors declare no conflict of interest.

References

1. Beavers, D.P.; Beavers, K.M.; Miller, M.; Stamey, J.; Messina, M.J. Exposure to isoflavone-containing soy products and endothelial function: A Bayesian meta-analysis of randomized controlled trials. *Nutr. Metab. Cardiovasc. Dis.* **2012**, *22*, 182–191. [CrossRef] [PubMed]
2. Taku, K.; Melby, M.K.; Takebayashi, J.; Mizuno, S.; Ishimi, Y.; Omaori, T.; Watanabe, S. Effect of Soy isoflavone extract supplements on bone mineral density in menopausal women: Meta-analysis of randomized controlled trials. *Asia Pac. J. Clin. Nutr.* **2010**, *19*, 33–42. [PubMed]
3. Balanos, R.; Cstillo, D.; Anelica, F.F. Soy isoflavones versus placebo in the treatment of climacteric vasomotor symptoms: systematic review and meta-analysis. *Menopause* **2010**, *17*, 660–666.
4. Li, Q.; Kubota, C. Effects of supplemental light quality on growth and phytochemicals of baby leaf lettuce. *Environ. Exp. Bot.* **2009**, *67*, 59–64. [CrossRef]
5. Samuoliene, G.; Brazaityte, A.; Sirtautas, R.; Sakalauskiene, S.; Jankauskiene, J.; Duchovskis, P.; Novičkovas, A. The impact of supplementary short-term red LED lighting on the antioxidant properties of microgreens. *Acta Hortic.* **2012**, *956*, 649–655. [CrossRef]
6. Vaštakaitė, V.; Viršilė, A.; Brazaitytė, A.; Samuolienė, G.; Jankauskienė, J.; Sirtautas, R.; Novičkovas, A.; Dabašinskas, L.; Sakalauskienė, S.; Miliauskienė, J.; et al. The Effect of Blue Light Dosage on Growth and Antioxidant Properties of Microgreens. *Sodinink. Daržinink.* **2015**, *34*, 25–35.
7. Sun, J.; Nishio, J.N.; Vogelmann, T.C. Green light drives CO_2 fixation deep within leaves. *Plant Cell Physiol.* **1998**, *39*, 1020–1026. [CrossRef]
8. Cui, J.; Ma, Z.H.; Xu, Z.G.; Zgang, H.; Chang, T.T.; Liu, H.J. Effects of supplemental lighting with different light qualities on growth and physiological characteristics of cucumber, pepper and tomato seedlings. *Acta Hortic. Sin.* **2009**, *5*, 663–670.
9. Swartz, T.E.; Corchnoy, S.B.; Christie, J.M.; Lewis, J.W.; Szundi, I.; Briggs, W.R. The photocycle of a flavin-binding domain of the blue light photoreceptor phototropin. *J. Biol. Chem.* **2001**, *276*, 36493–36500. [CrossRef] [PubMed]
10. Baroli, I.; Price, G.D.; Badger, M.R.; Von Caemmerer, S. The contribution of photosynthesis to the red light response of stomatal conductance. *Plant Physiol.* **2008**, *146*, 737–747. [CrossRef] [PubMed]
11. Hogewoning, S.W.; Trouwborst, G.; Maljaars, H.; Poorter, H.; van Ieperen, W.; Harbinson, J. Blue light dose–responses of leaf photosynthesis, morphology, and chemical composition of *cucumis sativus* grown under different combinations of red and blue light. *J. Exp. Bot.* **2010**, *61*, 3107–3117. [CrossRef] [PubMed]
12. Eom, S.H.; Park, H.J.; Seo, D.W.; Kim, W.W.; Cho, D.H. Stimulating effects of far infra-red ray radiation on the release of antioxidative phenolics in grape berries. *Food Sci. Biotechnol.* **2009**, *18*, 362–366.

13. Wink, M. Modes of Action of Herbal Medicines and Plant Secondary Metabolites. *Medicines* **2015**, *2*, 251–286. [CrossRef] [PubMed]
14. Eom, S.H.; Jin, C.W.; Park, H.J.; Kim, E.H.; Chung, I.M.; Kim, M.J.; Yu, C.Y.; Cho, D.H. Far infrared ray irradiation stimulates antioxidant activity in vitis flexuosa thumb Berries. *Korean J. Med. Crop Sci.* **2007**, *15*, 319–323.
15. Jeong, S.M.; Kim, S.Y.; Kim, D.R.; Jo, S.C.; Nam, K.C.; Ahn, D.U. Effect of heat treatment on antioxidant activity of citrus peels. *J. Agric. Food Chem.* **2004**, *52*, 3389–3393. [CrossRef] [PubMed]
16. Lee, S.; Kim, J.; Jeong, S.; Kim, D.; Ha, J.; Nam, K.; Ahn, D. Effect of far-infrared radiation on the antioxidant activity of rice hulls. *J. Agric. Food Chem.* **2003**, *51*, 4400–4403. [CrossRef] [PubMed]
17. Samarakoon, K.; Senevirathne, M.; Lee, W.W.; Kim, Y.T.; Kim, J.I.; Oh, M.C.; Jeon, Y.J. Antibacterial effect of citrus press-cakes dried by high speed and far infrared radiation dryng methods. *Nutr. Res. Pract.* **2012**, *6*, 187–194. [CrossRef] [PubMed]
18. Ghimeray, A.K.; Sharma, P.; Hu, W.; Cheng, W.; Park, C.H.; Rho, H.S.; Cho, D.H. Far infrared assisted conversion of isoflavones and its effect on total phenolics and antioxidant activity in black soybean seed. *J. Med. Plants Res.* **2013**, *7*, 1129–1137.
19. Singleton, V.L.; Rossi, J.A. Colorimetry of total phenolics with phosphomolybdic-phosphotungstic acid reagents. *Am. J. Enol. Vitic.* **1965**, *16*, 144–458.
20. Braca, A.; Fico, G.; Morelli, I.; Simone, F.; Tome, F.; Tommasi, N. Antioxidant and free radical scavenging activity of flavonol glycosides from different Aconitum species. *J. Ethnopharmacol.* **2003**, *8*, 63–67. [CrossRef]
21. Benzie, I.F.F.; Strain, J.J. The Ferric Reducing Ability of Plasma (FRAP) as a measure of "Antioxidant Power": The FRAP assay. *Anal. Biochem.* **1996**, *239*, 70–76. [CrossRef] [PubMed]
22. Lee, S.J.; Ahn, J.K.; Khanh, T.D.; Chun, S.C.; Kim, S.L.; Ro, H.M.; Song, H.K.; Chung, I.M. Comparison of Isoflavone Concentrations in Soybean (*Glycine max* (L.) Merrill) Sprouts Grown under Two Different Light Conditions. *J. Agric. Food Chem.* **2007**, *55*, 9415–9421. [CrossRef] [PubMed]
23. Meng, X.; Xing, T.; Wang, X. The role of light in the regulation of anthocyanin accumulation in Gerbera hybrid. *Plant Growth Regul.* **2004**, *44*, 243–250. [CrossRef]
24. Cevallos-Casals, B.A.; Cisneros-Zevallos, L. Impact of germination on phenolic content and antioxidant activity of 13 edible seed species. *Food Chem.* **2010**, *119*, 1485–1490. [CrossRef]
25. Lee, J.D.; Shannon, J.G.; Jeong, Y.S.; Lee, J.M.; Hwang, Y.H. A simple method for evaluation of sprout characters in soybean. *Euphytica* **2007**, *153*, 171–180. [CrossRef]
26. Randhir, R.; Kwon, Y.I.; Shetty, K. Effect of thermal processing on phenolic, antioxidant activity and health functionality of select grain sprouts and seedlings. *Innov. Food Sci. Emerg. Technol.* **2008**, *9*, 355–364. [CrossRef]
27. López-Amorós, M.L.; Hernández, T.; Estrella, I. Effect of germination on legume phenolic compounds and their antioxidant activity. *J. Food Comp. Anal.* **2006**, *19*, 277–283. [CrossRef]
28. Narukawa, Y.; Ichikawa, M.; Sanga, D.; Sano, M.; Mukai, T. White light emitting diodes with super high luminous efficacy. *J. Phys. Appl. Phys.* **2010**, *43*, 354002. [CrossRef]
29. Wu, M.C.; Hou, C.Y.; Jiang, C.M.; Wang, Y.T.; Wang, C.Y.; Chen, H.H. A novel approach of LED light radiation improves the antioxidant activity of pea seedlings. *Food Chem.* **2007**, *101*, 1753–1758. [CrossRef]
30. Samuoliene, G.; Urbonaviciute, A.; Brazaityte, A.; Sabajeviene, G.; Sakalauskaite, J.; Duchovskis, P. The impact of LED illumination on antioxidant protertise of sprouted seeds. *Cent. Eur. J. Biol.* **2011**, *6*, 68–74.

© 2018 by the authors. Licensee MDPI, Basel, Switzerland. This article is an open access article distributed under the terms and conditions of the Creative Commons Attribution (CC BY) license (http://creativecommons.org/licenses/by/4.0/).

Article

Characterization of Sparkling Wines According to Polyphenolic Profiles Obtained by HPLC-UV/Vis and Principal Component Analysis

Anaïs Izquierdo-Llopart * and Javier Saurina *

Department of Chemical Engineering and Analytical Chemistry, University of Barcelona, Martí i Franquès 1-11, 08028 Barcelona, Spain
* Correspondence: anais.izquierdo.llopart@gmail.com (A.I.-L.); xavi.saurina@ub.edu (J.S.); Tel.: +34-934021277 (J.S.)

Received: 13 December 2018; Accepted: 8 January 2019; Published: 10 January 2019

Abstract: Cava is a sparkling wine obtained by a secondary fermentation in its own bottle. Grape skin contains several compounds, such as polyphenols, which act like natural protectors and provide flavor and color to the wines. In this paper, a previously optimized method based on reversed phase high performance liquid chromatography (HPLC) with ultraviolet/visible (UV/Vis) detection has been applied to determine polyphenols in cava wines. Compounds have been separated in a C_{18} core-shell column using 0.1% formic acid aqueous solution and methanol as the components of the mobile phase. Chromatograms have been recorded at 280, 310 and 370 nm to gain information on the composition of benzoic acids, hidroxycinnamic acids and flavonoids, respectively. HPLC-UV/vis data consisting of compositional profiles of relevant analytes has been exploited to characterize cava wines produced from different base wine blends using chemometrics. Other oenological variables, such as vintage, aging or malolatic fermentation, have been fixed over all the samples to avoid their influence on the description. Principal component analysis and other statistic methods have been used to extract of the underlying information, providing an excellent discrimination of samples according to grape varieties and coupages.

Keywords: liquid chromatography; polyphenols; protected designation of origin; coupages; sparkling wine (cava); characterization; chemometrics

1. Introduction

Cava is a sparkling wine of high quality with Protected Designation of Origin (PDO) produced by the Champenoise method based on the second fermentation and aging period in its own bottle [1,2]. Cava is highly popular in our society: it is currently the most exported Spanish wine. For this reason, control strategies are needed to guarantee the standards of quality and the maintenance of organoleptic characteristics among winemaking batches. Part of this evaluation is carried out by sensory analysis with a group of expert panelists. Alternatively, analytical methods can be implemented to satisfy the increasing demand of controls, to reduce costs and to achieve reliable results [3]. The study of polyphenols in wines is a hot topic in the field of food analysis, because of the implications on organoleptic and descriptive issues [3,4]. For instance, the role of anthocyanins as principal pigments of wine or the influence of tannins on bitterness and astringency are well-known, while, apparently, they have a little direct contribution to odor [5].

Red wines from black skin grapes are much richer in polyphenols than rosé or white wines due to the simple fact that these compounds are extracted from the skin grape during the maceration process. Furthermore, cavas produced from the Chardonnay variety have a high content in polyphenolic substances; however, they differ in composition from those found in the varieties of black grapes [6].

Moreover, polyphenols are the substrate for some enzymatic changes of must and the main responsible for non-enzymatic auto-oxidation processes of the wines. Polyphenols have also shown a great interest as nutraceuticals because of the wide range of beneficial effects attributed, including antioxidant, antimicrobial and anti-inflammatory activities [7,8].

It has been pointed out elsewhere that naturally occurring components of food product may serve as efficient descriptors of some food features such as origin, varietal constituents, manufacturing practices, etc. [3,9]. Among the different food matrices, wines have been studied extensively on the basis of contents of chemical species, such as volatile compounds [10], inorganic elements [11,12], organic acids [13], amino acids and biogenic amines [14], polyphenols [15,16], etc. The corresponding compositional profiles and instrumental fingerprints have been used as the source of chemical information [17] to tackle characterization, classification and authentication issues, often with the assistance of chemometric methods.

In the field of cava and wines, polyphenolic data can be correlated with factors (e.g., geographic origin, climate and terrain, grape varieties, time of harvest and winemaking procedures) with a high impact on quality and economic issues, such as the qualification with the PDO or the final price [6]. Some applications have been reported on the role of polyphenols and related compounds as descriptors of cava features. It should be remarked that, in some of the cases mentioned below, principal component analysis (PCA) and related chemometric methods (cluster analysis, discriminant analysis, etc.) were used to facilitate the extraction of relevant information. A pioneering publication evaluated the descriptive ability of main components of base white wines from Macabeu, Xarel·lo and Parellada grapes to infer on their discrimination according to varieties, vintages or wineries; in the study, authors pointed out the possibilities of species such as cinnamic acids as potential varietal markers [18]. In another case, Martinez-Lapuente et al. studied the effect of alternative grape varieties on the polyphenolic composition and sensorial features of sparkling wines [19]. Patters on color and taste attributes and overall quality depending on varieties were drawn. It was concluded that Alvaring and Verdejo wines resulted in highly attractive alternatives. In a recent paper, the effect of some oenological practices on the composition of phytochemicals of rosé sparkling wines was evaluated; the influence of some additives to preserve the levels of the most abundant polyphenols was assessed as well [20]. Stefenon and coworkers investigated the influence of aging on lees on the polyphenolic content and antioxidant activity of sparkling wines obtained under Champenoise and Charmat winemaking protocols [21]. The authors found out changes in the evolution of the contents of species, such as tirosol, gallic acid, resveratrol or piceid, which varied differently depending on the vinification method. Another study compared some parameters of Champenoise, Charmat and Asti methods on the aromatic composition of Moscato Giallo sparkling wines. The authors concluded that the total polyphenolic content and the in vitro antioxidant activity of samples did not show significant differences among the three vinifications [22]. An investigation on biological aging and storage of cava wines revealed differences in the composition of browning components related to phenolic and furfural species as a function of the process time; the concentration of hydroxylmethylfurfural could be proposed as an index of cava quality [23]. A similar research also demonstrated the influence of oenological and aging processes on color changes and phenolic composition of rosé sparkling wines; in this case, color features could be associated to the transformation of polymeric species into absorbing anthocyanins as a consequence of aging [24]. Finally, Bosch-Fusté et al. investigated the viability of the total index of phenols defined as the absorbance at 280 nm as a quality marker of cava wines subjected to an accelerated aging process. The study of two sets of samples showed an increase in the phenolic index after seven weeks of accelerated breeding [25].

In this paper, relevant polyphenols were determined in different cava samples by an HPLC-UV/vis method previously established [26]. Compounds were separated chromatographically on a C_{18} column under an elution program based on increasing the percentage of methanol. In order to focus on varietal and blending issues more specifically, samples under study were chosen homogeneously regarding some winemaking factors such as vintage, aging in cellars and application of malolactic fermentation. As a result, descriptors of wine blending could be investigated without the influence of other disturbances. PCA was applied to compositional data to carry out an exploratory study of the coupages made with different grape varieties. Despite the lack of selective descriptors, results showed that various polyphenols were more abundant in some coupages. As a result, we concluded that the proposed method resulted in a simple, fast and suitable approach towards the characterization and classification of cavas according to base wine varieties and blends using polyphenolic concentrations as the source of information.

2. Materials and Methods

2.1. Chemicals and Solutions

The mobile phase was prepared with formic acid (>96%, Sigma-Aldrich, St. Louis, MO, USA), methanol (UHPLC-Supergradient, Panreac, Barcelona, Spain) and water (Elix3, Millipore, Bedford, MA, USA). Polyphenols of analytical quality were purchased from Sigma-Aldrich (St. Louis, MO, USA): gallic, homogentisic, protocatechuic, caftaric, gentisic, vanillic, caffeic, syringic, ferulic, *p*- coumaric acids, (+)-catechin, (−)-epicatechin, ethyl gallate, resveratrol, rutin, myricetin and quercetin. These compounds were selected according to their abundance and relevance in white and rosé cavas. Stock standard solutions of 1000 mg L^{-1} of each polyphenol were prepared in MeOH. Working standard solutions consisting of a mixture of analytes at concentrations ranging from 20 to 0.05 mg L^{-1} were prepared in MeOH:water (1:1, *v:v*).

2.2. Samples

Cava samples under analysis were kindly provided by the winery Codorníu Raventós S.A. Cavas were produced from base wines of Penedès and Costers del Segre regions (both from Catalonia, Spain). Samples consisted of white and rosé cavas all of them of 2015 vintage and aged for a period of 18 months, elaborated from various coupages of base wines of Chardonnay, Macabeu, Xarel·lo, Parellada, Pinot Noir, black Garnacha and Trepat varieties as follows: Coupage C (58 samples) corresponded to the three classical varieties: Macabeu, Xarel·lo and Parellada; Coupage E (six samples) was based on the classical combination with a small percentage of Chardonnay; Coupage A (15 samples) consisted of Chardonnay (70%) plus Macabeu, Xarel·lo and Parellada blending (30%); Coupage G (three samples) was Chardonnay 100%; Coupage S (two samples) was Chardonnay (50%) plus Macabeu and Xarel·lo (50%); Coupage T (four samples) corresponded to a rosé cava from Chardonnay (30%) and Pinot Noir (70%); Coupage V (nine samples) corresponded to a rosé cava from Pinot Noir, black Garnacha and Trepat. For each coupage, samples were independently produced (i.e., from different wine batches) and bottled. From the point of view of sugar content, the set comprised 11 brut nature (<3 g L^{-1} sugar), 61 brut (3–12 g L^{-1}), 13 dry (17–32 g L^{-1}) and 12 semi-dry (32–50 g L^{-1}).

Samples were degasified and filtered through a nylon membrane (0.45 μm pore size) prior HPLC-UV/vis analysis. A quality control solution consisting of mixture of 50 μL of each cava sample was prepared to evaluate the reproducibility of the chromatographic method and the significance of the descriptive PCA models. Cava samples were analyzed randomly and the quality control was injected every 10 samples.

2.3. Chromatographic Method

The chromatographic system consisted of an Agilent Series 1100 HPLC Chromatograph (Agilent Technologies, Palo Alto, CA, USA) equipped with a quaternary pump (G1311A), a degasser (G1322A), an automatic injection system (G1392A) and a diode array detector (G1315B). An Agilent ChemStation for LC 3D (Rev. A. 10.02) software was used for instrument control and data processing.

The chromatographic method was optimized and validated elsewhere [26]. Briefly, the separation was carried out in a Kinetex C_{18} column (Phenomenex, Torrance, CA, 100 mm × 4.6 mm internal diameter with 2.6 μm particle size). 0.1% of formic acid in water (*v/v*) (solvent A) and methanol (solvent B) were used to create the following elution gradient and the flow rate was 1 mL min^{-1}: from 0 to 20 min, B(%) 15–60 lineal increase; from 20 to 22 min, B(%) 60–90 lineal increase; from 22 to 27 min, B(%) 90 isocratic range, cleaning step; from 27 to 27.5 min, B(%) 90–15; from 27.5 to 30, B(%) 15 isocratic range, conditioning step. The injection volume was 10 µL. Chromatograms were acquired at 280, 310 and 370 nm.

2.4. Data Analysis

Preliminary statistics on polyphenolic contents in cava samples were gained from box and radial plots obtained with Excel. Exploratory studies by Principal Component Analysis (PCA) were carried out using the PLS-Toolbox working with MATLAB. The data matrix to be treated consisted of concentration values of quantified polyphenols in the cava samples. Data was autoscaled to achieve a similar weight to all the polyphenols regardless differences in amplitude and magnitude.

Sample patterns related to the different grape varieties and blends and other oenological practices were deduced form the interpretation of the scatter plot of scores of the first principal components (PCs). The distribution of variables on the space of the PC, shown in the plot of loadings, provided information on polyphenol correlations. Compounds up- or down expressed, which could be considered as tentative markers, were assessed from the simultaneous interpretation of scores and loadings graphs.

3. Results and Discussion

As indicated above, an HPLC-UV/vis method previously developed for the determination of relevant polyphenols in white wines [26] was here extended to cava (sparkling) wines. Samples under study were similar with respect to various oenological features such as aging, vintage, malolactic fermentation or Champenoise vinification, thus, the variability was associated to the varietal origin of grapes and the combination of base wines. The three main cava classes corresponded to the so-called classical blend (Macabeu, Xarel.lo and Parellada), the rosé blend (Pinot Noir, black Garnacha and Trepat) and the Chardonnay monovarietal cava. The other cava types evaluated consisted of combinations of classical, rosé and Chardonnay wines in different percentages. Figure 1 shows the chromatograms of representative rosé, classical and Chardonnay samples recorded at 280 nm. The chromatogram of a standard mixture of polyphenol at 5 mg L^{-1} was also included for identification purposes. It can be seen that most of the selected compounds were detected although some differences in the concentration levels were encountered among coupages. Analytes were quantified by linear regression models stablish using 10 standard mixtures with concentrations ranging from 0.05 to 20 mg L^{-1}. Differences in compositional profiles were the basis of further sample discrimination as a function of base wine varieties and coupages.

The analysis of the 97 samples under study revealed that, in general, hydroxycinnamic acids and their esters with tartaric acid were predominant species in this kind of winemaking process; benzoic acids were also abundant, while flavonoids were scarcer. As detailed in Table 1, wine samples were rich in gentisic and caftaric acids, with mean concentrations of 32.2 and 11.2 mg L^{-1}, respectively. Gallic, homogentisic and caffeic acids, and catechin occurred at moderate levels, ranging from 1 to 5.5 mg L^{-1}. Other minor components such as epicatechin, protocatechuic acid, p-coumaric acid vanillic acid and syringic acid were found at mean concentrations below 1 mg L^{-1}. Resveratrol and flavonoids, such as rutin and myricetin, were scarce (not detected in all the samples). The descriptive ability of each polyphenol (i.e., the information contained in each variable) was accounted from the relative standard deviation of the concentration in the series of cavas. The variability ranges given in Table 1 indicated that concentrations of gallic and gentisic acids were quite homogeneous in all the samples regardless of grape varieties or coupages, with overall relative standard deviation (RSD) values below 10%. This finding suggested that the apparent discriminant capacity of these polyphenols to classify cava wines was certainly limited. In contrast, contents of compounds such as epicatechin, protocatechuic acid, vanillic acid, ferulic acid and syringic acid were more heterogeneous, with RSD (%) values higher than 30%. As a result, we guessed that they could result in biomarker candidates to distinguish the different cava classes.

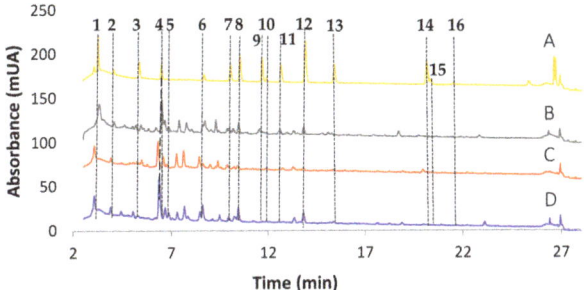

Figure 1. Chromatograms recorded at 280 nm using the proposed HPLC-UV/vis method. (**A**) Standard solution at 5 mg L^{-1}; (**B**) Rosé wine composed of Pinot Noir, Trepat and black Garnacha varieties (**C**) Classical grape varieties coupage composed of Macabeu, Xarel·lo and Parellada and (**D**) White cava composed of Chardonnay variety. Peaks assignment: (1) gallic acid, (2) homogentisic acid, (3) protocatechuic acid, (4) caftaric acid, (5) gentisic acid, (6) catechin, (7) vanillic acid, (8) caffeic acid, (9) syringic acid, (10) ethyl gallate, (11) epicatechin, (12) *p*-coumaric acid, (13) ferulic acid, (14) resveratrol, (15) rutin and (16) myricetin.

Table 1. Average concentration values of polyphenols in the set of samples under study.

Compounds	Average Concentration (mg L^{-1})	SD	RSD (%)
Gallic acid	4.1	0.3	7.0
Homogentisic acid	5.2	0.8	15.5
Protocatechuic acid	0.7	0.2	31.4
Caftaric acid	11.2	1.2	13.6
Gentisic acid	32.2	3.2	9.7
Catechin	4.5	0.8	28.1
Caffeic acid	1.5	0.2	14.5
p-Coumaric acid	0.74	0.13	19.3
Vanillic acid	0.92	0.21	46.5
Syringic acid	0.34	0.16	76.9
Epicatechin	0.60	0.21	66.8
Ferulic acid	0.12	0.14	120
Resveratrol	0.04	0.07	170
Rutin	0.007	0.035	165
Myricetin	0.029	0.089	92.0

Standard deviation (SD) and relative standard deviation (RSD) indicated the variability of concentrations as a measure of discriminating capacity among samples.

Boxplots and radial diagrams were used to explore the performance of polyphenols to distinguish cava as a function of coupages. Figure 2 depicts the overall analyte content and mean concentration values of analytes in the white and rosé cava classes.

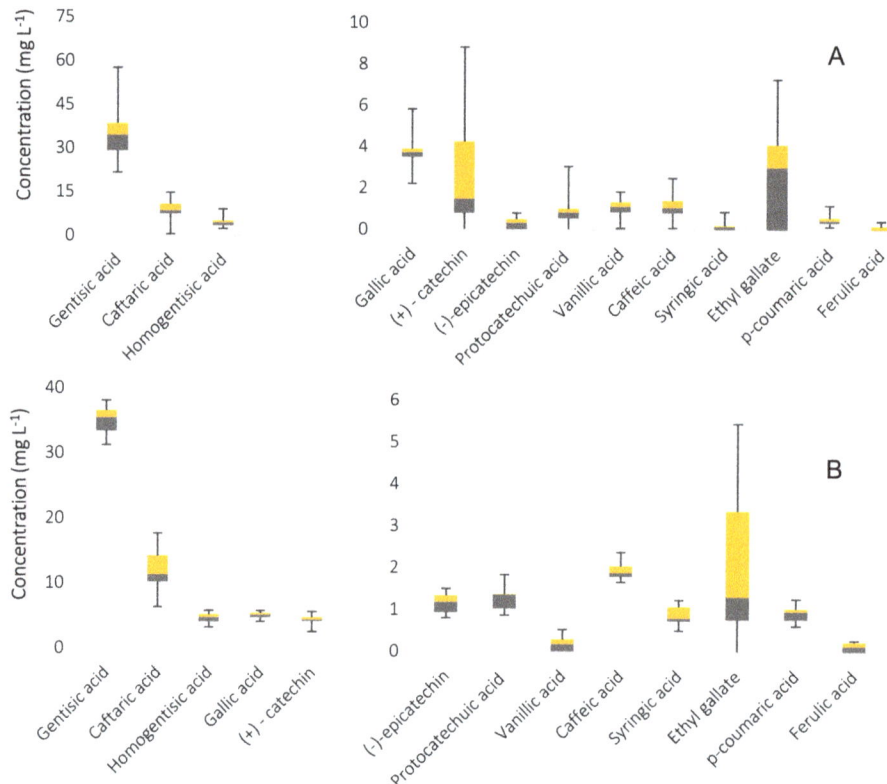

Figure 2. Boxplots with whiskers with the polyphenolic composition of the sets of (**A**) white and (**B**) rosé cava samples under study. Error bars indicated the variability in the concentration values.

In general, rosé (Pinot Noir, Trepat and black Garnacha combinations) and Chardonnay cavas showed the richest overall polyphenol contents, while the classical bends of Macabeu, Xarel·lo and Parellada cavas contained lower polyphenol levels (Figure 3A). In accordance with our previous results, gentisic acid was homogenously distributed among classes (Figure 3B); concentrations of the vast majority of samples ranged between 25 and 35 mg L^{-1}. This compound was not up- or down-expressed in any particular variety, thus it resulted in a poor descriptor. Additionally, gallic acid was widespread, although its concentration was slightly increased in rosé samples with Pinot Noir and black Garnacha varieties. Syringic acid was important in rosé cavas while it was quite residual in classical and Chardonnay-based cavas (Figure 3C); a similar pattern was observed for protocatechuic acid and epicatechin. Catechin (Figure 3D) and homogentisic acid were more abundant in Chardonnay and occurred at low concentrations in the coupages rich in classical varieties; as a result, catechin contents were highly dependent on the percentages of this variety in the samples. Another interesting pattern corresponded to vanillic acid which was present at low concentrations in rosé cavas (ca. 0.2 mg L^{-1}) and reached 5- to 10-fold higher concentrations in the other classes (Figure 3E). All the hydroxycinnamic acids (caffeic, ferulic, coumaric, etc.) behaved in a similar way, being more abundant in Chardonnay and rosé samples. Other compounds were less relevant for descriptive purposes as they displayed a less systematic behavior.

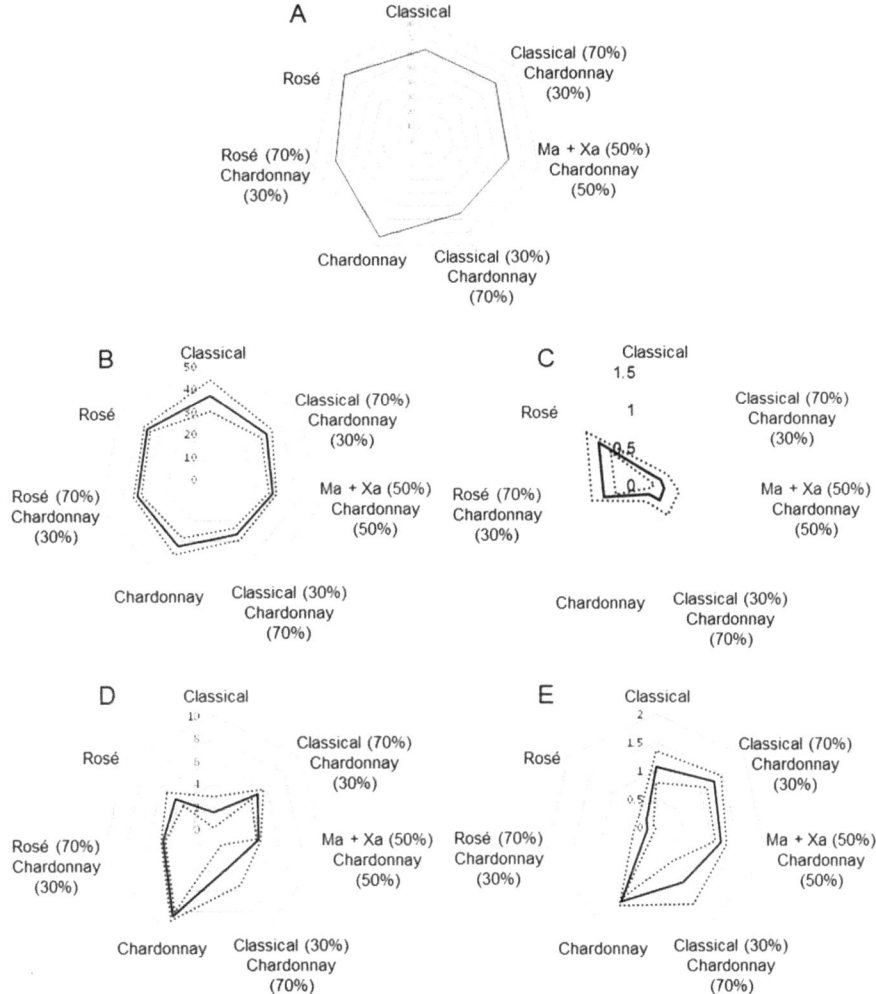

Figure 3. Radial plots of polyphenolic concentrations in the different coupages. (**A**) Overall content of analytes; (**B**) gentisic acid; (**C**) syringic acid; (**D**) catechin; (**E**) vanillic acid. Solid line indicates the mean value; dotted lines indicated the ± standard deviation values. Variety assignation: Ma, Macabeu; Xa, Xarel·lo.

Principal Component Analysis

Compositional profiles were used as the data to be treated by PCA for a more comprehensive study focused on identifying potential markers of the different classes of wine varieties and blends. The dataset consisted of concentration values of each polyphenol in the cava samples and the quality controls (QCs). As remarked in the experimental section, QCs were included to assess the soundness of the model. A first exploration carried out on the whole data matrix indicated that samples were reasonably distributed according to the coupages. Some minor components, such as resveratrol, rutin and myricetin, with concentrations close to the detection limits of the method, displayed high variability (RSD values in the QC replicates higher than 20%); thus, they were excluded from the dataset to obtain a more robust and accurate description of sample behavior. The rest of the variables, which exhibited RSD values lower that 6%, were taken for analysis.

The PCA model working with the refined set indicated that circa 75% of the information was captured with three principal components (PCs). In particular, PC1 retained 41.3% of variance, PC2 23.3% and PC3 9.9%. Scatter plots of scores of the relevant PCs revealed interesting patterns on sample distribution. In the case of PC1 versus PC2 (Figure 4A), QCs were located in a compact group in the central area of the graph, thus proving that the overall model was reliable and robust.

Interestingly, PC1 modeled the overall polyphenolic content with the richest cavas on the right and the poorest on the left. PC2 explained differences among rosé, classical and Chardonnay (top, center and bottom, respectively). More specifically, three main trends in the sample distribution were encountered, namely: (i) cavas in the left side were mainly elaborated with classical grape varieties of Macabeu, Xarel·lo and Parellada; (ii) samples in the bottom-right corner corresponded to monovarietal Chardonnay; (iii) rosé cavas prepared with black grapes appeared in the top-right corner. Intermediate situations of coupages elaborated with variable percentages of Chardonnay (e.g., classical varieties combined with 30, 50 or 70% of Chardonnay or rosé varieties combined with 30% Chardonnay) were distributed accordingly (see Figure 4A). Finally, although not shown here, PC3 mainly modeled the variations of Chardonnay proportions in the coupages, with larger scores for those samples containing higher percentages.

The scatter plot of loadings of PC1 versus PC2 (Figure 4B) described the behavior of the polyphenolic variables. It was first deduced that the levels of some compounds were reasonably correlated, e.g., syringic and gallic acids, and caffeic and coumaric acids, with correlation coefficients higher than 0.6. Other species were more negatively correlated (e.g., ethyl gallate and caftaric acid). The simultaneous study of scores and loadings indicated that classical cava coupages contained higher amounts of gentisic acid and ethyl gallate as they occupied the left side of the graphs. Syringic and gallic acids were representative of rosé cavas elaborated with red grapes varieties of Pinot Noir, Trepat and red Garnacha. Finally, catechin, homogentisic, caftaric, caffeic and coumaric acids were more abundant in Chardonnay cavas and coupages created with high percentages of this variety.

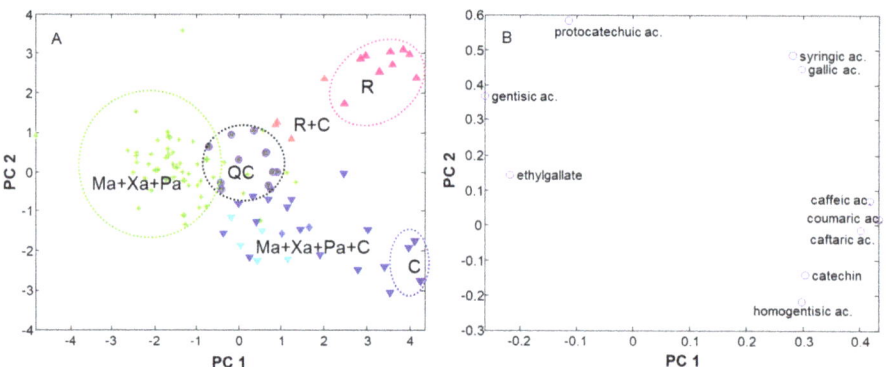

Figure 4. Principal component analysis of the dataset consisting of polyphenol concentrations of each cava sample. (**A**) Plot of scores of PC1 versus PC2; (**B**) plot of loadings of PC1 versus PC2. Acronyms: C Chardonnay; Ma Macabeu; Pa Parellada; Xa Xarel·lo; QC Quality control; R rosé. Symbols: Star = classical coupage (Ma + Xa + Pa); Triangle (vertex up) = rosé cava; Triangle (vertex down) = Chardonnay cava; circle = QC.

Apart from the relationship between phenolic matter and grape varieties, the sugar content was another variable of the set of samples under study. The liquor of expedition after bottle disgorging provided different concentrations of sugar, ranging from 50 g L^{-1} for semi-dry to less than 3 g L^{-1} for brut nature. In the current case, PCA plots showed that the sugar content was uncorrelated with polyphenols and had no influence on the sample distribution. Regarding geographical factors and agricultural practices that may affect substantially the phenolic levels, their effect on the set of samples under study was buffered, since enologists combined appropriate proportions of base wines from different regions to obtain products with more constant physicochemical and sensorial attributes, according to commercial parameters of quality and the costumer's preferences. Finally, as indicated above, other variables such as vintage and aging, which may notably influence on the polyphenolic composition, were controlled for; thus, patterns observed mainly depended on the combinations of base wines.

4. Conclusions

Sparkling wines belonging to the protected designation of origin Cava were analyzed with a chromatographic method for the determination of polyphenols. Wines from different coupages were considered including monovarietal Chardonnay samples, rosé samples composed of Pinot Noir, Trepat and black Garnacha, the classical blends of Macabeu, Xarel·lo and Parellada and other mixtures containing variable amounts of Chardonnay. Other oenological variables, such as vintage, aging and malolactic fermentation, were fixed for all the samples so that they did not affect the description. In general, overall polyphenolic contents were higher in Chardonnay wines, followed by rosé, while the classical coupages were circa 20% poorer in these components. Patterns deduced from boxplots and radial diagrams were confirmed from a comprehensive data exploration by principal component analysis. Although no molecule was found to be a specific descriptor of cava classes, syringic acid was more abundant in rosé wines, the hydroxycinnamic acids were up-expressed in Chardonnay samples and vanillic acid was almost residual in red grape products. Summarizing, polyphenolic profiling combined with a chemometric data analysis resulted in an excellent approach to deal with the characterization of cava wines according to the coupages. This strategy could be extended to further classification and authentication purposes.

Author Contributions: A.I.-L. performed all the experiments and wrote the paper. J.S. conceived, supervised the work and revised the paper.

Funding: This research was funded by the Spanish Ministry of Economy and Competitiveness (project CTQ2014-56324-C2-1-P) and by Generalitat de Catalunya (project 2017 SGR-171).

Acknowledgments: The authors gratefully acknowledge the support received from and Codorníu Raventós winery providing the samples.

Conflicts of Interest: The authors declare no conflict of interest.

References

1. Buxaderas, S.; Lopez-Tamames, E. Sparkling Wines: Features and Trends from Tradition. *Adv. Food Nutr. Res.* **2012**, *66*, 1–45. [PubMed]
2. Institut del Cava. Available online: http://www.institutdelcava.com/en/ (accessed on 10 January 2019).
3. Saurina, J. Characterization of wines using compositional profiles and chemometrics. *Trend Anal. Chem.* **2010**, *29*, 234–245. [CrossRef]
4. Herderich, M.J.; Smith, P.A. Analysis of grape and wine tannins: Methods, applications and challenges. *Aust. J. Grape Wine Res.* **2005**, *11*, 205–214. [CrossRef]
5. Welch, C.R.; Wu, Q.L.; Simon, J.E. Recent advances in anthocyanin analysis and characterization. *Curr. Anal. Chem.* **2008**, *4*, 75–101. [CrossRef] [PubMed]
6. Linskens, H.F.; Jackson, F. *Wine Analysis*; Springer: Berlin, Germany, 1988; Volume 6.
7. Bravo, L. Polyphenols: Chemistry, dietary sources, metabolism, and nutritional significance. *Nutr. Rev.* **1998**, *56*, 317–333. [CrossRef] [PubMed]
8. Yao, L.H.; Jiang, Y.M.; Shi, J.; Tomas-Barberan, F.A.; Datta, N.; Singanusong, R.; Chen, S.S. Flavonoids in food and their health benefits. *Plant Food Hum. Nutr.* **2004**, *59*, 113–122. [CrossRef]
9. Ignat, I.; Volf, I.; Popa, V.I. A critical review of methods for characterisation of polyphenolic compounds in fruits and vegetables. *Food Chem.* **2011**, *126*, 1821–1835. [CrossRef]
10. Torrens, J.; Riu-Aumatell, M.; Vichi, S.; Lopez-Tamames, E.; Buxaderas, S. Assessment of Volatile and Sensory Profiles between Base and Sparkling Wines. *J. Agric. Food Chem.* **2010**, *58*, 2455–2461. [CrossRef]

11. Jos, A.; Moreno, I.; Gonzalez, A.G.; Repetto, G.; Camean, A.M. Differentiation of sparkling wines (cava and champagne) according to their mineral content. *Talanta* **2004**, *63*, 377–382. [CrossRef]
12. Geana, E.I.; Marinescu, A.; Iordache, A.M.; Sandru, C.; Ionete, R.E.; Bala, C. Differentiation of Romanian Wines on Geographical Origin and Wine Variety by Elemental Composition and Phenolic Components. *Food Anal. Methods* **2014**, *7*, 2064–2074. [CrossRef]
13. Huang, X.Y.; Jiang, Z.T.; Tan, J.; Li, R. Geographical Origin Traceability of Red Wines Based on Chemometric Classification via Organic Acid Profiles. *J. Food Qual.* **2017**, *2017*, 2038073. [CrossRef]
14. Garcia-Villar, N.; Hernandez-Cassou, S.; Saurina, J. Characterization of wines through the biogenic amine contents using chromatographic techniques and chemometric data analysis. *J. Agric. Food Chem.* **2007**, *55*, 7453–7461. [CrossRef] [PubMed]
15. Franquet-Griell, H.; Checa, A.; Nunez, O.; Saurina, J.; Hernandez-Cassou, S.; Puignou, L. Determination of Polyphenols in Spanish Wines by Capillary Zone Electrophoresis. Application to Wine Characterization by Using Chemometrics. *J. Agric. Food Chem.* **2012**, *60*, 8340–8349. [CrossRef] [PubMed]
16. Serrano-Lourido, D.; Saurina, J.; Hernandez-Cassou, S.; Checa, A. Classification and characterisation of Spanish red wines according to their appellation of origin based on chromatographic profiles and chemometric data analysis. *Food Chem.* **2012**, *135*, 1425–1431. [CrossRef] [PubMed]
17. Diaz, R.; Gallart-Ayala, H.; Sancho, J.V.; Nunez, O.; Zamora, T.; Martins, C.P.B.; Hernandez, F.; Hernandez-Cassou, S.; Saurina, J.; Checa, A. Told through the wine: A liquid chromatography-mass spectrometry interplatform comparison reveals the influence of the global approach on the final annotated metabolites in non-targeted metabolomics. *J. Chromatogr. A* **2016**, *1433*, 90–97. [CrossRef] [PubMed]
18. De La Presa Owens, C.; Lamuela Raventos, R.M.; Buxaderas, S.; De La Torre Boronat, M.C. Characterization of Macabeo, Xarello, and Parellada white wines from the Penedes region. II. *Am. J. Enol. Vitic.* **1995**, *46*, 529–541.
19. Martinez-Lapuente, L.; Guadalupe, Z.; Ayestaran, B.; Ortega-Heras, M.; Perez-Magarino, S. Sparkling Wines Produced from Alternative Varieties: Sensory Attributes and Evolution of Phenolics during Winemaking and Aging. *Am. J. Enol. Vitic.* **2013**, *64*, 39–49. [CrossRef]
20. Sartor, S.; Burin, V.M.; Panceri, C.P.; dos Passos, R.R.; Caliari, V.; Bordignon-Luiz, M.T. Rose Sparkling Wines: Influence of Winemaking Practices on the Phytochemical Polyphenol during Aging on Lees and Commercial Storage. *J. Food Sci.* **2018**, *83*, 2790–2801. [CrossRef]
21. Stefenon, C.A.; Bonesi, C.D.; Marzarotto, V.; Barnabe, D.; Spinelli, F.R.; Webber, V.; Vanderlinde, R. Phenolic composition and antioxidant activity in sparkling wines: Modulation by the ageing on lees. *Food Chem.* **2014**, *145*, 292–299. [CrossRef]
22. Caliari, V.; Panceri, C.P.; Rosier, J.P.; Bordignon-Luiz, M.T. Effect of the Traditional, Charmat and Asti method production on the volatile composition of Moscato Giallo sparkling wines. *LWT Food Sci. Technol.* **2015**, *5*, 393–400. [CrossRef]
23. Serra-Cayuela, A.; Aguilera-Curiel, M.A.; Riu-Aumatell, M.; Buxaderas, S.; Lopez-Tamames, E. Browning during biological aging and commercial storage of Cava sparkling wine and the use of 5-HMF as a quality marker. *Food Res. Int.* **2013**, *53*, 226–231. [CrossRef]
24. Torchio, F.; Segade, S.R.; Gerbi, V.; Cagnasso, E.; Rolle, L. Changes in chromatic characteristics and phenolic composition during winemaking and shelf-life of two types of red sweet sparkling wines. *Food Res. Int.* **2011**, *44*, 729–738. [CrossRef]
25. Bosch-Fuste, J.; Sartini, E.; Flores-Rubio, C.; Caixach, J.; Lopez-Tamames, E.; Buxaderas, S. Viability of total phenol index value as quality marker of sparkling wines, "cavas". *Food. Chem.* **2009**, *114*, 782–790. [CrossRef]
26. Larrauri, A.; Nunez, O.; Hernandez-Cassou, S.; Saurina, J. Determination of Polyphenols in White Wines by Liquid Chromatography: Application to the Characterization of Alella (Catalonia, Spain) Wines Using Chemometric Methods. *J. AOAC Int.* **2017**, *100*, 323–329. [CrossRef] [PubMed]

© 2019 by the authors. Licensee MDPI, Basel, Switzerland. This article is an open access article distributed under the terms and conditions of the Creative Commons Attribution (CC BY) license (http://creativecommons.org/licenses/by/4.0/).

Article

Characterization and Determination of Interesterification Markers (Triacylglycerol Regioisomers) in Confectionery Oils by Liquid Chromatography-Mass Spectrometry

Valentina Santoro [1,*], Federica Dal Bello [1], Riccardo Aigotti [1], Daniela Gastaldi [1], Francesco Romaniello [1,2], Emanuele Forte [2], Martina Magni [2], Claudio Baiocchi [1] and Claudio Medana [1]

1. Department of Molecular Biotechnology and Health Sciences, University of Turin, Turin 10124, Italy; federica.dalbello@unito.it (F.D.B.); riccardo.aigotti@unito.it (R.A.); daniela.gastaldi@unito.it (D.G.); francesco.romaniello@unito.it (F.R.); claudio.baiocchi@unito.it (C.B.); claudio.medana@unito.it (C.M.)
2. Soremartec Italy srl (Ferrero Group), AlbaCN 12051, Italy; Emanuele.FORTE@ferrero.com (E.F.); Martina.MAGNI@ferrero.com (M.M.)
* Correspondence: valentina.santoro@unito.it; Tel.: +39-011-670-5240

Received: 15 January 2018; Accepted: 12 February 2018; Published: 16 February 2018

Abstract: Interesterification is an industrial transformation process aiming to change the physico-chemical properties of vegetable oils by redistributing fatty acid position within the original constituent of the triglycerides. In the confectionery industry, controlling formation degree of positional isomers is important in order to obtain fats with the desired properties. Silver ion HPLC (High Performance Liquid Chromatography) is the analytical technique usually adopted to separate triglycerides (TAGs) having different unsaturation degrees. However, separation of TAG positional isomers is a challenge when the number of double bonds is the same and the only difference is in their position within the triglyceride molecule. The TAG positional isomers involved in the present work have a structural specificity that require a separation method tailored to the needs of confectionery industry. The aim of this work was to obtain a chromatographic resolution that might allow reliable qualitative and quantitative evaluation of TAG positional isomers within reasonably rapid retention times and robust in respect of repeatability and reproducibility. The resulting analytical procedure was applied both to confectionery raw materials and final products.

Keywords: confectionery fats; positional isomers; silver-ion HPLC; triglycerides

1. Introduction

There is a great demand from food industry of fats with different degrees of solidity. However, most of the fats coming from natural sources are oils so they need to be suitably transformed in order to modulate their hardness. Interesterification, by changing the position of fatty acids within the original triglyceride structure, is the process more commonly used to modify consistency of oils [1]. Such result may be obtained either chemically using a catalyst or by the use of enzymes. In both cases, positional isomers (regioisomers) from original triglycerides are formed and the amount of their formation modulates the final physical properties of the fat.

TAGs (triglycerides) composition of a fat is perceived by the consumer since it is associated with the spreading properties, hardness and palatability of the product.

Small variations in TAG structure may have an impact on crystallization and polymorphic properties of fats. Thermal processes applied for technological reasons (e.g., deodorization) may produce positional isomers, inducing changes in the physical characteristics of fats. In fact, deodorization at high temperatures induces formation of asymmetric TAGs influencing crystallization behavior. The above changes must be monitored by analytical methods and kept under control in the course of the industrial process in relation to quality control and research and development of new TAGs composition that can improve product quality.

Among natural fats, cocoa butter has a particular relevance because of its unusual and highly valued physical properties and its peculiar triglycerides' content. They are symmetrical monounsaturated TAGs containing three major fatty acids: palmitic (P), stearic (S) and oleic (O) and three main glycerides, palmitoyl-oleoyl-palmitoyl glycerol (POP), palmitoyl-oleoyl-stearoyl glycerol (POS) and stearoyl-oleoyl-stearoyl glycerol (SOS), with the unsaturated fatty acid located in the central position as it is typical in vegetable oils.

If thermal process for deodorization is applied to cocoa butter, positional isomers may be produced with a change in the physical characteristics of the product. As already said, deodorization at high temperatures induces a significant formation of asymmetric TAGs influencing the crystallization behavior.

The main analytical technique for regiospecific analysis of TAGs is silver-ion high performance liquid chromatography (HPLC) coupled via an atmospheric pressure chemical ionization (APCI) source to a mass (MS) detector. Silver ion chromatography is based on the interaction of silver ions, bounded to a cation-exchange stationary phase, with π electrons of double bonds of unsaturated TAGs [2,3]. However, even if silver-ion HPLC remains the unique separation technique of unsaturated fats, its intrinsically difficult use in terms of repeatability and reproducibility must be underlined. Generally, the equilibration times are long and separations in gradient conditions are not easily reproducible.

In addition, conditioning times between successive analyses are time-consuming due to the necessary transition between two mobile phase compositions having different strength.

Furthermore, separation in isocratic conditions does require daily, before starting a sequence, long conditioning times and blank chromatographic runs to attain reproducibility. Separation of TAGs containing fatty acids having different unsaturation degrees by silver-ion HPLC has been the object of several publications [4–18].The aim of many of them was to obtain a detailed qualitative profile of complex mixtures of animal and vegetable fats. However, baseline separation of variously unsaturated TAGs, necessary for reliable quantitative analyses, was not always achieved because of the matrix complexity. In many cases, when three columns in series were used, retention times were inevitably long [15,16].

More recently, the use of traditional reversed phase column (polymeric bonded phase) was reported. However, the separation of critical regioisomer couples not always was enough to assure accurate quantitative evaluation of each component and their retention times were also long [19].

An interesting recent work [20] presents a rapid separation of TAG regioisomers in animal fats by Differential Mobility Spectrometry (DMS). However, the optimization of various experimental parameters such as the separation and compensation voltages applied to DMS electrodes, the type and flow rate of chemical modifier and the dwell time of analyte ions in the DMS cell is required. In addition, to be expensive, the technique is sophisticated and scarcely suitable to the analytical skill generally present in industry internal quality control laboratory.

The aim of the present paper is to describe a separation method of individuated typical TAGs of cocoa butter, palm and other tropical oils mainly used in confectionery industry. The main couples of triacylglycerol regioisomers that had to be separated, identified and quantified were, respectively, POP/PPO (palmitoyl-palmitoyl-oleoyl glycerol), POS/PSO (palmitoyl-stearoyl-oleoyl glycerol, SOS/SSO (stearoyl-stearoyl-oleoyl glycerol) and POO/OPO (palmitoyl-oleoyl-oleoyl glycerol/oleoyl-palmitoyl-oleoyl glycerol), SOO/OSO (stearoyl-oleoyl-oleoyl glycerol/oleoyl-stearoyl-oleoyl glycerol).

The chromatographic problem we had to face up to was challenging because we needed to obtain retention times as low as possible and selective enough to allow a reliable peak integration. A task not easy to obtain because we had two classes of compounds with very different retention times due to a different number of double bonds (e.g., POP in respect to POO, etc.) and with a very subtle structural difference within the single class (different position of an oleic acid residue within the triglyceride structure). Separation of the regioisomer couples before cited, reported in literature, was characterized by very long retention times (40–60 min) [14–16] not compatible with the analysis times useful for a confectionery industry internal control laboratory.

As a matter of fact, the high chromatographic selectivity necessary to separate the regioisomers belonging to the same class is conflicting with the quite lower selectivity required to separate the constituents of the two classes. In other words, higher selectivity conditions (elution phase of low force) necessary to separate the couples POP/PPO, POS/PSO and SOS/SSO would cause the species POO/OPO and SOO/OSO to have very different and longer retention times. As a consequence, we will try to find a compromise and, differently from the works reported in literature, we used cation exchange columns silver modified by us in order to achieve a better control on retention mechanism by varying, if necessary, the degree of silver modification of the stationary phase.

Due to the small number of TAG isomers to be separated belonging to two classes of molecules differing by the number of double bonds, we adopted two different isocratic steps and a rapid transition between them as a more controllable separation mode.

2. Materials and Methods

2.1. Reagents

Isopropanol, acetonitrile, n-heptane and ethyl acetate LC-MS (Liquid Chromatography-Mass Spectrometry)-grade solvents, POP, PPO, POS, PSO, SOS, SSO, OPO, POO, SOO, and OSO TAG pure standards and anhydrous Na_2SO_4 were purchased from Sigma-Aldrich (St. Louis, MO, USA).

2.2. Sample Preparation

Samples obtained from confectionery industry were previously heated to complete fusion and homogenized by mechanical stirring. Successively were dehydrated with Na_2SO_4 and filtered through Buchner. A sample aliquot (10.0 mg) was added to of n-heptane (10 mL) and analyzed after suitable dilution.

2.3. Calibration Curve Preparation

A stock solution of all standards mixed together was prepared in n-heptane at a concentration of 1000 ppm each. To draw a calibration curve, the stock solution was diluted to contain 1.0, 2.0, 3.0, 5.0, 8.0, 10.0 ppm, respectively.

2.4. Instrumentation

The separation system was a binary solvent HPLC Ultimate 3000 (Thermo Fisher Scientific, Milan, Italy) interfaced through an APCI ionization source to a linear ion trap coupled to a high resolution mass analyzer (LTQ-Orbitrap Thermo Fisher Scientific, Milan, Italy). APCI ionization source heated at 450 °C was used in positive ions mode in the mass range 240–1000 Th.

Capillary temperature was 250 °C, flow rate of sheath gas and auxiliary gas were set at 35.0 and 15.0 arbitrary units, capillary voltage was 25.0 V, source voltage 6.0 kV, and tube lens 110.0 V. Mass resolution was set at 30,000.

2.5. Chromatography

A cation exchange column (Luna SCX, Phenomenex, 150 × 2.0 mm, 5 µm, 100 Å) silver-modified by us was used. A gradient separation made up by two isocratic steps at different mobile phase compositions separated by a rapid transition time of five minutes between them was used. The two mobile phase compositions (n-heptane:ethylacetate 93:7 and 90:10, respectively) were chosen according to elution times and peak resolution. Flow rate was 0.300 mL/min. The injection volume was 10.0 µL.

2.6. Silver Modification of SCX Columns

After the inversion of the flow direction, we rinsed the column with methanol/acetonitrile (CH_3OH/CH_3CN) 1:1 for about 45–60 min (flow rate 1.0 mL/min). As the concentration of $AgNO_3$ solution and the number and volume of injections determine the retention characteristic of the modified SCX stationary phase, we tested three concentrations of $AgNO_3$ solution (0.57 M, 1.18 M and 1.86 M). During the rinsing procedure, ten 50.0 µL injections of a 1.18 M $AgNO_3$ solution in CH_3CN allowed for obtaining a chromatographic result suited to our requirements. The column was successively inverted to the original position and rinsed with dichloromethane for an hour at a flow rate of 0.20 mL/min.

2.7. Statistical Evaluation of Final Data and Validation

The analytical procedure, the object of the present work, is destined to be applied to raw material and to final confectionery products (interesterified or not), so real samples of whatever concentration may be at one's disposal. However, validation parameters like Limit of Detection (LOD) or Limit of Quantitation (LOQ) were anyway evaluated. Very important for our purposes are parameters like linearity, measure uncertainty, repeatability and reproducibility. Their definition is assuming particular significance in the present analytical context because taking under control the mobile and stationary phase conditions ruling the chromatographic performances was a particularly demanding task as described before.

In addition, as already stated, a determination of the amount of positional isomers, as accurate as possible, is functional to establish a reliable relationship with the interesterification conditions.

A calibration curve was realized in six replicates and each curve was made by six points spanning from 1.0 to 10.0 ppm. The straight line was calculated by a linear regression method and each calibration point was the mean of the six replicates. The R^2 value was used as a first linearity evaluator. The successive statistical calculations were based on the evaluation of the mean calibration curve error. Hubaux-Vos formulae were applied [21].

In Table 1, all data concerning linearity, measure uncertainty, repeatability, reproducibility, LOD and LOQ for the positional isomers object of this study are reported along with the mean calibration curves for the standards.

Table 1. Validation data: calibration curve parameters, repeatability, reproducibility, uncertainty measure, limit of detection (LOQ) and limit of quantification (LOQ).

Tag	Calibration Points (ppm)	Calibration Curve Parameters [1]	Rsd (%)	Calibration Curve Parameters [2]	Rsd (%)	Measure Uncertainty $x_0 \pm t\,(0.05, 5)s_{x0}$	LOD, LOQ (ppm)
SOS	1	$b = (5.91 \pm 0.06) \times 10^6$ $R^2 = 0.9996$	13.9	$b = (5.82 \pm 0.10) \times 10^6$ $R^2 = 0.9985$	10.3	1.00 ± 0.38	0.31, 1.04
	2		6.73		8.85	1.89 ± 0.34	
	3		5.99		6.81	2.87 ± 0.32	
	5		6.39		5.09	5.01 ± 0.29	
	8		2.13		7.90	7.61 ± 0.34	
	10		1.22		2.25	10.37 ± 0.46	
SSO	1	$b = (7.79 \pm 0.07) \times 10^6$ $R^2 = 0.9996$	2.42	$b = (7.84 \pm 0.10) \times 10^6$ $R^2 = 0.9992$	10.6	0.97 ± 0.27	0.31, 1.03
	2		15.5		9.96	2.19 ± 0.24	
	3		6.46		5.65	2.91 ± 0.22	
	5		4.01		5.71	4.69 ± 0.21	
	8		1.82		2.30	8.09 ± 0.25	
	10		1.09		9.86	10.1 ± 0.31	
POS	1	$b = (5.76 \pm 0.07) \times 10^6$ $R^2 = 0.9995$	13.9	$b = (5.61 \pm 0.04) \times 10^6$ $R^2 = 0.9997$	9.56	1.06 ± 0.16	0.30, 1.01
	2		6.03		5.10	2.17 ± 0.14	
	3		2.12		4.28	3.13 ± 0.13	
	5		4.13		5.18	4.91 ± 0.12	
	8		1.69		4.28	7.97 ± 0.15	
	10		8.64		10.6	9.99 ± 0.18	

Table 1. Cont.

Tag	Calibration Points (ppm)	Calibration Curve Parameters [1]	Rsd (%)	Calibration Curve Parameters [2]	Rsd (%)	Measure Uncertainty $x_0 \pm t\,(0.05, 5)s_{x_0}$	LOD, LOQ (ppm)
PSO	1		5.22		9.86	1.02 ± 0.21	
	2		7.16		8.57	2.03 ± 0.19	
	3	$b = (9.34 \pm 0.10) \times 10^6$	3.55	$b = (9.41 \pm 0.09) \times 10^6$	4.41	2.82 ± 0.17	0.30, 1.00
	5	$R^2 = 0.9985$	0.45	$R^2 = 0.9995$	3.52	4.96 ± 0.16	
	8		3.64		3.78	7.85 ± 0.19	
	10		4.05		6.86	10.2 ± 0.25	
POP	1		7.76		9.02	0.95 ± 0.27	
	2		4.89		7.48	2.14 ± 0.24	
	3	$b = (3.69 \pm 0.06) \times 10^6$	5.11	$b = (3.61 \pm 0.05) \times 10^6$	11.1	3.16 ± 0.22	0.26, 0.85
	5	$R^2 = 0.9989$	9.37	$R^2 = 0.9992$	11.3	4.84 ± 0.21	
	8		4.73		5.28	7.99 ± 0.25	
	10		2.81		12.5	10.1 ± 0.32	
PPO	1		11.1		10.02	1.07 ± 0.11	
	2		1.97		4.38	2.02 ± 0.10	
	3	$b = (12.60 \pm 0.07) \times 10^6$	4.59	$b = (12.67 \pm 0.06) \times 10^6$	4.97	3.04 ± 0.09	0.32, 1.07
	5	$R^2 = 0.9998$	0.75	$R^2 = 0.9999$	4.59	4.87 ± 0.08	
	8		3.42		5.14	7.99 ± 0.10	
	10		5.38		9.52	10.1 ± 0.12	
SOO	1		6.10		9.10	1.19 ± 0.35	
	2		4.13		2.62	2.07 ± 0.32	
	3	$b = (16.19 \pm 0.21) \times 10^6$	2.29	$b = (16.26 \pm 0.27) \times 10^6$	2.13	2.91 ± 0.30	0.32, 1.08
	5	$R^2 = 0.9992$	0.71	$R^2 = 0.9960$	8.05	5.38 ± 0.28	
	8		1.73		1.80	8.10 ± 0.34	
	10		1.96		4.74	9.72 ± 0.40	
OSO	1		0.99		4.25	1.20 ± 0.22	
	2		2.94		1.95	2.14 ± 0.20	
	3	$b = (22.37 \pm 0.25) \times 10^6$	1.56	$b = (22.31 \pm 0.23) \times 10^6$	2.65	3.11 ± 0.18	0.15, 0.51
	5	$R^2 = 0.9994$	3.89	$R^2 = 0.9990$	5.58	4.90 ± 0.17	
	8		3.49		4.66	7.86 ± 0.20	
	10		0.45		3.79	10.1 ± 0.26	

[1] Repeatability; mean of three replicates; [2] reproducibility; mean of six replicates.

3. Results and Discussion

In literature, there are no papers devoted explicitly to specific regioisomers separation. They are concerning characterization of triacylglycerol composition in animal fats [13] or complex natural triacylglycerol mixture [15]. General works regarding the effect of temperature and mobile phase composition [12] on regioisomer separation involved triacylglycerol positional isomers very different from those envisaged in the present work. Anyway, when couples like POP/PPO or SOS/SSO, important in our case, were occasionally present in some of the reported separations in literature, they showed retention times very long (between 40 and 60 min) and the obtained chromatographic selectivity was due to the use of three columns in series. Such separation conditions are incompatible with the typical exigencies of industrial analytical laboratories. Thus, we had to reformulate the separation in a manner more tailored to the molecular characteristics of the specific TAG regioisomers of interest in confectionery industry and to the performance of their internal laboratories.

As a matter of fact, the choice of chromatographic separation conditions characterized by two isocratic steps reported in the Materials and Methods section was the result of a fine-tuning of the mobile phase strength. The mobile phase composition must be strong enough to assure reasonably short retention times and, at the same time, not too strong to level out the chromatographic resolution based on different stereochemistry of TAG positional isomers.

At first, we checked the difference in selectivity between two elution strength modifiers like acetonitrile and ethyl acetate, which have a very different elution force. Mixtures of *n*-hexane or *n*-heptane and a very small percent of acetonitrile (0.8–0.11%) provide good separations of unsaturated TAGs [3–5]. In these cases, acetonitrile concentration must be very low due to its high elution force and its poor miscibility with *n*-hexane or *n*-heptane. Such a mixture is unstable and tends to give chromatographic results not easily reproducible. The HPLC binary pump was unsatisfactory in the delivery of the mixed mobile phase because of its low mixing equilibrium; consequently, pre-mixing was necessary. Sometimes, in order to overcome this problem, isopropanol is used to intermediate the two immiscible solvents [11–18]. The equilibrium of the mixture was always precarious. Subtle differences in mixture preparation, temperature oscillations and different rates of unavoidable evaporation from the HPLC reservoirs of the three mobile phase components were all factors that often had a substantial influence on retention times. Maintaining time elution reproducibility was a challenge, and day-by-day chromatographic variations were recorded, making conditioning times progressively longer. All of these troubles are perfectly managed in academic studies where careful attention to difficult experimental conditions can be paid. On the contrary, in the case of internal laboratories of confectionery industry, a more easily applicable protocol would be at one's disposal.

Alternatively, the use of ethyl acetate as a modifier of *n*-heptane allowed for obtaining slightly better reproducible chromatography, conditioning times shorter and by far better results' reproducibility.

Different elution conditions were also tested and eventually a satisfactory chromatographic resolution was obtained with two successive isocratic segments involving *n*-heptane:ethyl acetate compositions of 93:7 and 90:10, respectively, with five minutes to change from the initial to final composition. In addition, only one column was used to obtain a satisfactory separation.

In Figure 1, the separation of standard samples of regioisomeric couples POP/PPO, POS/PSO, SOS/SSO, POO/OPO and SOO/OSO, at the same concentration (10.0 ppm), monitored by a linear ion trap is shown.

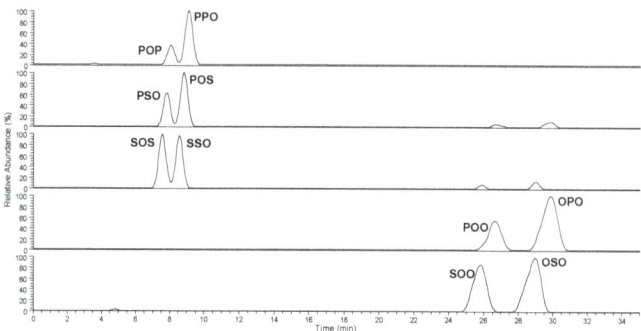

Figure 1. Chromatographic separation of standard TAG (triglyceride) regioisomers visualized at the extracted precursor ion m/z values.

As it can be seen, the chain length of TAG fatty acids has extremely limited influence on the retention times. The different spatial position of the double bond of oleic acid in sn-2 with respect to sn-1/sn-3 positions is enough to produce separation of regioisomers with the same number of double bonds whereas the presence of two double bonds change drastically the retention times.

The MS analyser provided structural information already coming from the thermal lability of triglycerides in the hot (450 °C) APCI ionization source. Triglycerides tend to release their constituent fatty acids in a typical way involving primarily the ones in the external position and to a lesser extent the one in the central position [22]. This behavior allows for distinguishing the isobaric couples of positional isomers.

An example of full mass spectra of standard TAG positional isomers POP/PPO is shown in Figure 2a,b. The qualitative pattern of thermal product ions is very similar but with different signal intensity as a consequence of the selective fragmentation pathway privileging the loss of fatty acids in external positions.

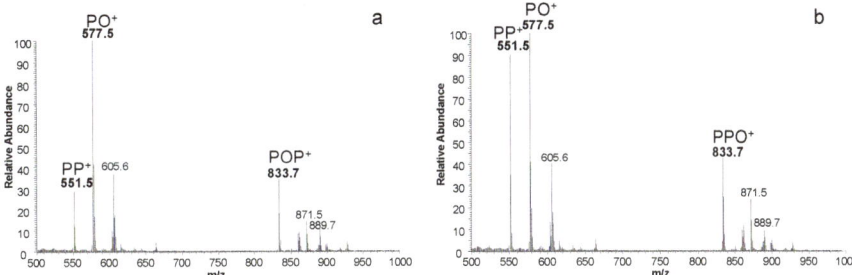

Figure 2. Full mass spectra of isobaric positional isomers POP (palmitoyl-oleoyl-palmitoyl glycerol) (**a**) and PPO (palmitoyl-palmitoyl-oleoyl glycerol) (**b**) showing the specificity of thermal fragmentation mechanism.

By applying the same experimental conditions to interesterified samples obtained from the confectionery industry, shea oil or palm olein, it is possible to obtain evidence of their interesterification degree, as shown in the chromatographic profiles of Figure 3a,b. As expected, the signal intensity of the different couples of regioisomers in the two oils reflects their TAG composition, in particular the low presence of palmitic acid in shea oil TAG constituents.

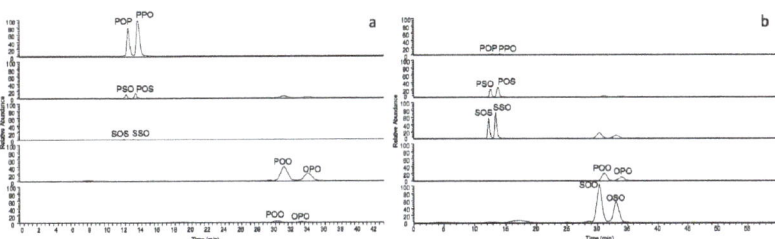

Figure 3. Chromatographic separation of TAG regioisomers typical of interesterified Shea oil (**a**) and Palm oil (**b**) visualized at the extracted precursor ion m/z values.

Figure 4a–c shows LC-MS chromatograms of a cocoa butter samples pre and after deodorization at two different temperatures (220 °C and 260 °C) corresponding to the increasing presence of thermally formed positional isomers, respectively (as an example, only the couple POP, PPO is reported).

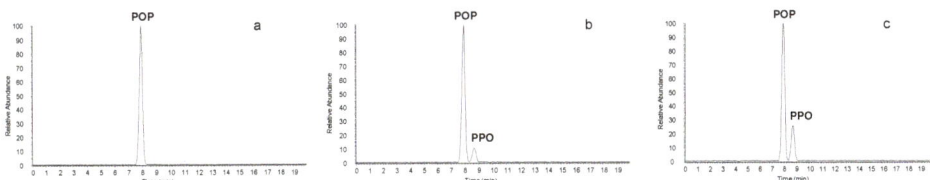

Figure 4. LC-MS (Liquid chromatography-mass spectrometry) chromatograms of cocoa butter samples pre deodorization (**a**) and deodorized at two different temperatures 220 °C (**b**) and 260 °C (**c**), corresponding to the absence and to the presence of thermally formed positional isomers, respectively.

In addition to confectionery industry raw materials, samples of final products were analyzed. In Tables 2–4, meaningful results are reported about a commercial biscuit filling cream and two spreading creams. As it can be seen, there is evidence of TAG regioisomer formation in two final products with respect to the original raw materials used.

Table 2. Comparison of TAG (triglyceride) percent composition between original raw materials and confectionery final products (biscuit filling cream).

TAG	Biscuit Filling Cream TAG Composition (%)	Original Raw Material	
		Shea Butter (%)	High Oleic Rapeseed Oil (%)
POP	84.1	91.51	100
PPO	15.9	8.49	0
POS	77.3	100	100
PSO	22.7	0	0
SOS	80.3	100	100
SSO	19.7	0	0
POO	100	100	100
OPO	0	0	0
SOO	88.2	100	100
OSO	11.8	0	0
	Evidence of interesterification		

Table 3. Comparison of TAG percent composition between original raw materials and confectionery final products (spreading cream A).

Tag	Spreading Cream A TAG Composition (%)	Original Raw Material	
		High Linoleic Sunflower Oil (%)	Cocoa Butter (%)
POP	95.7	100	100
PPO	4.3 [1]	0	0
POS	100	100	100
PSO	0	0	0
SOS	100	100	100
SSO	0	0	0
POO	100	100	100
OPO	0	0	0
SOO	100	100	100
OSO	0	0	0
	No interesterification		

[1] Coming from vaccine butter fraction present in milk powder.

Table 4. Comparison of TAG percent composition between original raw materials and confectionery final products (spreading cream B).

TAG	Spreading Cream BTAG Composition (%)	Original Raw Material	
		High Linoleic Sunflower Oil (%)	Cocoa Butter (%)
POP	83.0	100	100
PPO	17.0	0	0
POS	84.5	100	100
PSO	15.5	0	0
SOS	84.1	100	100
SSO	15.9	0	0
POO	91.7	100	100
OPO	8.3	0	0
SOO	91.7	100	100
OSO	8.3	0	0
	Evidence of interesterification		

Thus, it was possible to establish a reliable relationship between the extent of positional isomers formation and the experimental conditions to which industry samples can undergo. In this way, on the one hand, undesirable regioisomer formation may be avoided and, on the other, a regioisomer formation may be guided until the achievement of the desired physico-chemical properties.

4. Conclusions

Interesterification process of vegetable oils is of paramount importance in the confectionery industry because positional isomer formation allows for modulating the fat property. On the other hand, unwanted interesterification events may occur during thermal treatment of fats as in the deodorizing process. Therefore, either in the case of intentionally producing an interesterification process or checking its unwanted occurrence, it is necessary to have at one's disposal an analytical method helping to establish a sound relationship between the extent of positional isomer presence and the experimental conditions causing their formation.

In this work, we developed an analytical protocol based on LC-MS analysis able to identify and quantify the positional isomers typical of confectionery fats undergoing the interesterification process or thermal treatment. The time required for the global analysis was relatively short, the chromatographic resolution and efficiency were satisfactory and the mass detection allowed for identifying the isobaric components of each positional isomer couple.

In conclusion, the method described may well be considered a good diagnostic tool of interesterification consequences that are strictly connected to confectionery product quality.

Acknowledgments: The authors greatfully acknowledge Soremartec Italy srl for providing the raw materials for the experiments.

Author Contributions: V.S., F.D.B. and F.R. designed and performed the chemical analysis; V.S. and C.B. wrote the paper; D.G. and R.A. analyzed the data; E.F. and M.M. proposed the topic and contributed reagents/materials/analysis tools; C.M. performed a scientific supervision and manuscript revising.

Conflicts of Interest: The authors declare no conflict of interest.

References

1. Gupta, M.K. *Practical Guide to Vegetable Oil Processing*, 1st ed.; AOCS Press: Urbana, IL, USA, 2010; ISBN 978-1-89-399790-5.
2. De Vries, B. Quantitative separation of lipid materials by column chromatography on silica impregnated with silver nitrate. *Chem. Ind. J.* **1962**, *24*, 1049–1050.
3. Adlof, R.O.; Menzel, A.; Dorovska-Taran, V. Analysis of conjugated linoleic acid-enriched triacylglycerol mixtures by isocratic silver-ion high-performance liquid chromatography. *J. Chromatogr. A* **2002**, *953*, 293–297. [CrossRef]
4. Christie, W.W. Separation of molecular species of triacylglycerols by high-performance liquid chromatography with a silver ion column. *J. Chromatogr. A* **1988**, *454*, 273–284. [CrossRef]
5. Adlof, R.O. Analysis of triacylglycerol positional isomers by silver ion high-performance liquid chromatography. *J. High Resolut. Chromatogr.* **1995**, *18*, 105–107. [CrossRef]
6. Adlof, R.O. Normal-phase separation effects with lipids on a silver ion high-performance liquid chromatography column. *J. Chromatogr. A* **1997**, *764*, 337–340. [CrossRef]
7. Laakso, P.; Voutilainen, P. Analysis of triacylglycerols by silver-ion high-performance liquid chromatography—Atmospheric pressure chemical ionization mass spectrometry. *Lipids* **1996**, *31*, 1311–1322. [CrossRef] [PubMed]
8. Schuyl, P.J.W.; de Joode, T.; Vasconcellos, M.A.; Duchateau, G.S.M.J.E. Silver-phase high-performance liquid chromatography–electrospray mass spectrometry of triacylglycerols. *J. Chromatogr. A* **1998**, *810*, 53–61. [CrossRef]
9. Mondello, L.; Tranchida, P.Q.; Stanek, V.; Jandera, P.; Dugo, G.; Dugo, P. Silver-ion reversed-phase comprehensive two-dimensional liquid chromatography combined with mass spectrometric detection in lipidic food analysis. *J. Chromatogr. A* **2005**, *1086*, 91–98. [CrossRef] [PubMed]

10. Dugo, P.; Kumm, T.; Crupi, M.L.; Cotroneo, A.; Mondello, L. Comprehensive two-dimensional liquid chromatography combined with mass spectrometric detection in the analyses of triacylglycerols in natural lipidic matrixes. *J. Chromatogr. A* **2006**, *1112*, 269–275. [CrossRef] [PubMed]
11. Nikolova-Damyanova, B. Retention of lipids in silver ion high-performance liquid chromatography: Facts and assumptions. *J. Chromatogr. A* **2009**, *1216*, 1815–1824. [CrossRef] [PubMed]
12. Lisa, M.; Velínská, H.; Holčapek, M. Regioisomeric Characterization of Triacylglycerols Using Silver-Ion HPLC/MS and Randomization Synthesis of Standards. *Anal. Chem.* **2009**, *81*, 3903–3910. [CrossRef] [PubMed]
13. Lisa, M.; Netusilova, K.; Franek, L.; Dvorakova, H.; Vrkoslav, V.; Holcapek, M. Characterization of fatty acid and triacylglycerol composition in animal fats using silver-ion and non-aqueous reversed-phase high-performance liquid chromatography/mass spectrometry and gas chromatography/flame ionization detection. *J. Chromatogr. A* **2011**, *1218*, 7499–7510. [CrossRef] [PubMed]
14. Lisa, M.; Denev, R.; Holčapek, M. Retention behavior of isomeric triacylglycerols in silver-ion HPLC: Effects of mobile phase composition and temperature. *J. Sep. Sci.* **2013**, *36*, 2888–2900. [CrossRef] [PubMed]
15. Holčapek, M.; Velínská, H.; Lísa, M.; Česla, P. Orthogonality of silver-ion and non-aqueous reversed-phase HPLC/MS in the analysis of complex natural mixtures of triacylglycerols. *J. Sep. Sci.* **2009**, *32*, 3672–3680. [CrossRef] [PubMed]
16. Holčapek, M.; Dvorakova, H.; Lisa, M.; Giron, A.J.; Sandra, P.; Cvacka, J. Regioisomeric analysis of triacylglycerols using silver-ion liquid chromatography atmospheric pressure chemical ionization mass spectrometry: Comparison of five different mass analyzers. *J. Chromatogr. A* **2010**, *1217*, 8186–8194. [CrossRef] [PubMed]
17. Byrdwell, W.C.; Holčapek, M. *Extreme Chromatography: Faster, Hotter, Smaller*, 1st ed.; American Oil Chemists Society: Urbana, IL, USA, 2011; pp. 197–230. ISBN 978-1-89-399766-0.
18. Ovčačíková, M.; Lisa, M.; Cifkova, E.; Holcapek, M. Retention behavior of lipids in reversed-phase ultrahigh-performance liquid chromatography-electrospray ionization mass spectrometry. *J. Chromatogr. A* **2016**, *1450*, 76–85. [CrossRef] [PubMed]
19. TambaSompila, A.W.G.; Héron, S.; Hmida, D.; Tchapla, A. Fast non-aqueous reversed-phase liquid chromatography separation of triacylglycerol regioisomers with isocratic mobile phase. Application to different oils and fats. *J. Chromatogr. B* **2017**, *1041–1042*, 151–157. [CrossRef] [PubMed]
20. Šala, M.; Lísa, M.; Campbell, J.L.; Holčapek, M. Determination of triacylglycerol regioisomers using differential mobility spectrometry. *Rapid Commun. Mass Spectrom.* **2016**, *30*, 256–264. [CrossRef] [PubMed]
21. Hubaux, A.; Vos, G. Decision and detection limits for linear calibration curves. *Anal. Chem.* **1970**, *42*, 849–855. [CrossRef]
22. Baiocchi, C.; Medana, C.; Dal Bello, F.; Giancotti, V.; Aigotti, R.; Gastaldi, D. Analysis of regioisomers of polyunsaturated triacylglycerols in marine matrices by HPLC/HRMS. *Food Chem.* **2015**, *166*, 551–560. [CrossRef] [PubMed]

© 2018 by the authors. Licensee MDPI, Basel, Switzerland. This article is an open access article distributed under the terms and conditions of the Creative Commons Attribution (CC BY) license (http://creativecommons.org/licenses/by/4.0/).

Article

Authentication and Quantitation of Fraud in Extra Virgin Olive Oils Based on HPLC-UV Fingerprinting and Multivariate Calibration

Núria Carranco [1], Mireia Farrés-Cebrián [1], Javier Saurina [1,2] and Oscar Núñez [1,2,3,*]

[1] Department of Chemical Engineering and Analytical Chemistry, University of Barcelona, Martí i Franqués, 1-11, E08028 Barcelona, Spain; nuria.carranco.cruz@gmail.com (N.C.); mirefarres28@gmail.com (M.F.-C.); xavi.saurina@ub.edu (J.S.)
[2] Research Institute in Food Nutrition and Food Safety, University of Barcelona, Recinte Torribera, Av. Prat de la Riba 171, Edifici de Recerca (Gaudí), Santa Coloma de Gramenet, E08921 Barcelona, Spain
[3] Serra Húnter Fellow, Generalitat de Catalunya, Rambla de Catalunya 19-21, E08007 Barcelona, Spain
* Correspondence: oscar.nunez@ub.edu; Tel.: +34-93-403-3706

Received: 30 January 2018; Accepted: 20 March 2018; Published: 21 March 2018

Abstract: High performance liquid chromatography method with ultra-violet detection (HPLC-UV) fingerprinting was applied for the analysis and characterization of olive oils, and was performed using a Zorbax Eclipse XDB-C8 reversed-phase column under gradient elution, employing 0.1% formic acid aqueous solution and methanol as mobile phase. More than 130 edible oils, including monovarietal extra-virgin olive oils (EVOOs) and other vegetable oils, were analyzed. Principal component analysis results showed a noticeable discrimination between olive oils and other vegetable oils using raw HPLC-UV chromatographic profiles as data descriptors. However, selected HPLC-UV chromatographic time-window segments were necessary to achieve discrimination among monovarietal EVOOs. Partial least square (PLS) regression was employed to tackle olive oil authentication of Arbequina EVOO adulterated with Picual EVOO, a refined olive oil, and sunflower oil. Highly satisfactory results were obtained after PLS analysis, with overall errors in the quantitation of adulteration in the Arbequina EVOO (minimum 2.5% adulterant) below 2.9%.

Keywords: high performance liquid chromatography; UV detection; multivariate calibration; food authentication; olive oils; fraud quantitation

1. Introduction

Olive oil is one of the main ingredients in the Mediterranean diet, and over the past few years its consumption has also spread outside the Mediterranean basin. This is due to both its particular taste and remarkable nutritional properties. In fact, virgin olive oil (VOO) is the primary fat source in the Mediterranean diet, and it is obtained either by mechanical or direct pressing of the pulp of the olive fruit (*Olea europaea* L.). The olives, after being crushed to form pomace, are homogenized and pressed. VOO is not subjected to any treatment other than washing, decantation, centrifugation, and filtration [1]. Extra-virgin olive oil (EVOO) is considered to be the highest quality olive oil, and is characterized by high levels of beneficial constituents [2]. Apart from the natural unrefined state, olive oils of lower quality known as refined olive oils (ROOs) are also produced. For example, an extracted oil with poor sensory characteristics, high free fatty acid content, or peroxide values exceeding the established limits, is unfit for human consumption. This oil is then classified as lampante oil and needs to be refined. Additionally, pomace oils are solvent-extracted from olive pomace and also undergo refining [1,3]. Triglycerides are the major constituents of EVOOs, representing more than 98% of the total weight. The fatty acid composition of these triglycerides involves saturated fatty acids (11%), around 80% monounsaturated fatty acids (oleic acid being the most important feature of olive oils in comparison to other vegetable oils), and polyunsaturated fatty acids (9%). Furthermore, EVOO also contains antioxidant and anti-inflammatory compounds such as squalene, phytosterols, and highly bioavailable polyphenols and phenolic acids belonging to different families [2,4–9]. These compounds provide important beneficial health effects.

The beneficial characteristics of EVOOs and their high quality with respect to other edible oils make them an expensive product, susceptible to being adulterated with oils produced from cheaper fruits or seeds [10], or even with other lower quality olive oils. Moreover, lately there have been suspicions among consumers and administrations that in some cases producers have labeled virgin olive oils or even refined ones as EVOOs in order to increase profits. A comprehensive academic food fraud study spanning 30 years in the Journal of Food Science showed that olive oil was the single most commonly referenced adulterated food of any type in scholarly articles from 1980 to 2010 [11]. This is a relevant problem not only because of the financial issues affecting olive oil-producing countries such as Spain or Italy, but also in terms of health and consumer protection concerns. However, identification of some fraud is especially difficult, or can be time- and solvent-consuming when using traditional methodologies for the identification of adulteration, e.g., when the cheaper oil is added to EVOO at a level lower than a certain percentage. Therefore, the reduction of the current levels of fraud in olive oils requires the constant development of new analytical techniques for the detection of possible adulteration. In this regard, the study of compositional differences among olive oils and other edible vegetable oils which could be used as adulterants (e.g., because of their low cost) has been the basis for authentication approaches.

The types of polyphenols and phenolic acids and their concentration levels in olive oils depend on multiple parameters such as climate conditions, water resources, growing area, cultivation techniques, soil management, and the degree of maturation of the olives. Moreover, the oil production method (malaxation and extraction) can also affect the final quantities of these compounds [12–15]. As a result, polyphenol and phenolic acid content can be exploited as a source of analytical data to establish vegetable oil product classifications and to achieve product authentication [16,17], thus preventing fraud by detecting and quantifying product adulteration.

Several analytical methodologies have been employed in the determination of polyphenols in olive oils [18,19]. One of the most critical points is the extraction procedure, which is frequently addressed by liquid–liquid extraction (LLE) [20–24] and solid phase extraction (SPE), mainly using diol-bonded SPE cartridges [8,22,24–28]. Capillary electrophoresis (CE), liquid chromatography (LC), and gas chromatography (GC) are among the separation techniques usually employed for the determination of polyphenols [18,19,29]. However, the necessity for a derivatization step prior to GC separation [30] and the low sensitivity of CE methods when on-column UV detection is employed [20,24,31] make LC the technique of choice nowadays for the determination of polyphenols and phenolic compounds in olive oils. With respect to detection, electrochemical detection (ECD) [26], fluorescence [32], ultra-violet (UV) detection [8,24,32–34], mass spectrometry (LC-MS) [35] or tandem mass spectrometry (LC-MS/MS) [24,27,28,36] are currently used. Recently, due to the complexity of edible oil samples and the high structural variability of polyphenols, liquid chromatography coupled to high resolution mass spectrometry (LC-HRMS) appears to be the best technique for the identification and determination of phenolic compounds in vegetable oils [22,25], although these techniques are still expensive for many food control laboratories. Lately, interest in the classification and characterization of olive oils with respect to olive varieties, as well as in the identification and quantitation of fraud to guarantee product authentication, has been increasing, and several studies devoted to these topics can be found in the literature [21,37–40].

The aim of the present work was to develop a simple, cheap, and reliable high performance liquid chromatography method with UV detection (HPLC-UV) for the generation of polyphenolic fingerprints in the classification and authentication of olive oils. Data corresponding to the HPLC-UV polyphenolic fingerprints recorded at different wavelengths were considered as a source of potential descriptors to be exploited for the characterization and classification of edible oils (olive, sunflower, soy, and corn oils) and olive oils from different olive cultivars (Arbequina, Picual, Hojiblanca, and Cornicabra) by exploratory principal component analysis (PCA). Finally, Arbequina EVOO was adulterated with different amounts (2–85%) of Picual EVOO, a commercial low quality ROO, and a sunflower oil. The HPLC-UV data was evaluated by means of partial least squares (PLS) regression for authentication purposes as well as for the quantification of adulterant content.

2. Materials and Methods

2.1. Chemicals and Standard Solutions

Unless specified, analytical grade reagents were used. Methanol (Chromosolv™ for HPLC, ≥99.9%), hexane, and formic acid (≥98%) were obtained from Sigma-Aldrich (St. Louis, MO, USA), and ethanol (absolute) from VWR International Eurolab S.L. (Barcelona, Spain).

Water was purified using an Elix 3 coupled to a Milli-Q system (Millipore, Bedford, MA, USA) and was filtered through a 0.22-μm nylon membrane integrated into the Milli-Q system.

2.2. Instrumentation

An Agilent 1100 Series HPLC instrument equipped with a G1311A quaternary pump, a G1379A degasser, a G1392A autosampler, a G1315B diode-array detector, and a computer with the Agilent Chemstation software (Rev. A 10.02), all from Agilent Technologies (Waldbronn, Germany), was employed to obtain the HPLC-UV chromatographic fingerprints for the PCA and PLS studies. Chromatographic separation was carried out in reversed-phase mode by using a Zorbax Eclipse XDB-C8 column (150 × 4.6 mm i.d., 5 μm particle size) also provided by Agilent Technologies. Formic acid (0.1%, v/v) aqueous solution (solvent A) and methanol (solvent B) were used as mobile phase to stablish the gradient elution as follows: 0–2 min at 10% B (initial conditions); 2–4.5 min linear gradient from 10% B to 25% B; 4.5–7 min at 25% B; 7–22 min linear gradient from 25% B to 90% B; 22–24 min at 90% B; 24–25 min back to initial conditions at 10% B; and 25–30 min at 10% B for column equilibration. A mobile phase flow-rate of 1 mL min^{-1} and an injection volume of 10 μL were employed. Photodiode array (PDA) acquisition from 190 to 600 nm was performed to register UV spectra and to guarantee peak purity when necessary. HPLC-UV fingerprints for PCA and PLS analysis were obtained by direct UV absorption detection at 257, 280, and 316 nm.

2.3. Samples and Sample Treatment

Two sets of vegetable oil samples belonging to different trademarks and purchased from markets in Barcelona (Spain) were used in this work. The first set of samples consisted of 72 vegetable oils distributed as follows: 47 olive oils, 16 sunflower oils, 2 corn oils, 2 soy oils, and 5 vegetable oils produced from mixtures of seeds (3 sunflower/corn oils and 2 sunflower/soy oils). Among the 47 olive oils belonging to this first set of samples, 10 were obtained from Arbequina olives and 5 from Picual olives. No information regarding the olive cultivar of other 32 olive oil samples was available. The second set of samples consisted of 66 EVOO samples as follows: 23 from Arbequina olives, 19 from Picual olives, 12 from Hojiblanca olives, and 12 from Cornicabra olives. Additionally, several refined olive oil samples were also employed for the adulteration studies.

Sample treatment was carried out following a previously described method with some modifications [28]. For that purpose, 2.00 g of oil were weighed into a 15 mL polytetrafluoroethylene (PTFE) tube (Serviquimia, Barcelona, Spain), and extracted with 2 mL of an ethanol:water (70:30, v/v) solution by shaking vigorously for 2 min using a vortex (Stuart, Stone, UK). The extract was then centrifuged for 5 min at 3500 rpm (Rotanta 460 RS centrifuge, Hettich, Tuttlingen, Germany). In order to facilitate the quantitative recovery of the extract solution, the PTFE tube was then frozen for 24 h at $-18\ °C$, and the corresponding extract transferred into another 15 mL PTFE tube for further clean-up. Defatting was performed with 2 mL of hexane, shaking vigorously for 2 min in a vortex, and centrifuging for 5 min at 3500 rpm. Finally, the aqueous ethanolic extracts were transferred into 2 mL injection vials to be analyzed with the proposed HPLC-UV method.

Besides, a quality control (QC) consisting of a mixture of 50 µL of each ethanolic aqueous sample extract was prepared to evaluate the repeatability of the method and the robustness of the chemometric results.

For authentication studies by PLS regression, three cases were considered in which an Arbequina EVOO sample was adulterated with different amounts (from 2.5 to 85%) of a Picual EVOO sample, a ROO sample, or a sunflower oil sample, respectively. Hence, apart from those pure extracts (three samples each), mixtures of the Arbequina EVOO sample and the adulterant oil were as follows: 85% adulterant (3 samples), 80% adulterant (3 samples), 60% adulterant (3 samples), 55% adulterant (3 samples), 50% adulterant (8 samples), 40% adulterant (3 samples), 30% adulterant (3 samples), 20% adulterant (3 samples), 12% adulterant (3 samples), 10% adulterant (3 samples), 7% adulterant (3 samples), 5% adulterant (3 samples), 3% adulterant (3 samples), and 2.5% adulterant (3 samples), for each adulterant oil employed.

2.4. Data Analysis

SOLO chemometric software from Eigenvector Research was used for calculations with PCA and PLS regression [41]. A detailed description of the theoretical background of these methods is given elsewhere [42].

Data matrices to be treated by PCA consisted of the HPLC-UV chromatographic fingerprints obtained at different acquisition wavelengths (257, 280, and 316 nm). HPLC-UV chromatograms were pretreated to improve the data quality while minimizing solvent and matrix interferences, peak shifting, and baseline drifts (for additional details see [43]). In some cases, segmented HPLC-UV fingerprints were employed to improve PCA classification. Scatter plots of scores and loadings of the principal components (PCs) were used to investigate the structure of maps of samples and variables, respectively.

The percentage of the oil used for adulteration (Picual EVOO, ROO, or sunflower oil) in the adulterated Arbequina EVOO samples analyzed was quantified using PLS. Samples available were distributed among training and test sets as follows. Training set: 100% adulterant (3 samples), 80% adulterant (3 samples), 60% adulterant (3 samples), 50% adulterant (8 samples), 40% adulterant (3 samples), 20% adulterant (3 samples), 10% adulterant (3 samples), 5% adulterant (3 samples), 2.5% adulterant (3 samples), and 100% Arbequina EVVO (3 samples). The remaining samples considered as unknown (85% adulterant, 55% adulterant, 30% adulterant, 12% adulterant, 7% adulterant and 3% adulterant, 3 samples each) were used for validation and prediction purposes. For both training and test steps, X-data matrices consisted of the HPLC-UV chromatographic fingerprints and the Y-data matrices contained the oil adulteration percentages.

3. Results and Discussion

3.1. Exploratory Studies by Principal Component Analysis

As a first study, the characterization and classification of olive oils with respect to other edible vegetable oils (sunflower, corn, and soy oils, as well as mixtures of them) was attempted by using raw HPLC-UV chromatographic profiles (i.e., absorbance over time) as analytical data for PCA.

Reversed-phase HPLC-UV chromatographic fingerprints were obtained by employing a previously developed method in a Zorbax Eclipse XDB-C8 column under gradient elution using methanol and 0.1% aqueous formic acid solutions as mobile phase [21]. HPLC-UV chromatographic fingerprints were studied at three acquisition wavelengths: 257, 280, and 316 nm. The reproducibility of the extraction step and its influence on the PCA results was studied elsewhere [21]. Relative standard deviation (RSD%) values of polyphenol features ranged from 0.3 to 8.3% depending on the levels of components. Replicate samples have similar PCA scores, thus indicating that extraction variabilities did not significantly affect the descriptive and quantitative performance of chemometric models. The variability of chromatographic runs was evaluated from the repetitive injections of the QC (every 10 sample injections) throughout the series. In this way, the influence of peak shifting and/or baseline drift on the PCA models was deduced from trends observed in the scores plot. When QCs were not clustered randomly due to systematic deviations, corrective data pretreatments such as peak alignment and baseline subtraction were applied [43].

A first set of samples consisting of 72 commercially available vegetable oils was employed. The ethanolic extracts as well as the QCs (see Experimental section) were randomly analyzed with the proposed method, and the obtained HPLC-UV fingerprints subjected to PCA. At first, the performance of the resulting models was quite limited due to experimental variability of chromatograms, displaying peak shifting and baseline drifts that affected the distribution of samples. For this reason, chromatogram pre-processing was recommendable to improve the analytical performance of PCA models. Thus, data matrices were autoscaled and then subjected to PCA. Figure 1 shows the score plots of PC1 vs. PC2 when processing HPLC-UV chromatographic fingerprint data registered at (a) 257 nm; (b) 280 nm; and (c) 316 nm.

As can be seen, in general QCs appeared in compact groups, demonstrating a good repeatability and robustness of results, with the exception of the dataset at 316 nm (Figure 1c) where a relatively higher dispersion on QCs was observed. Regarding sample characterization and classification, a good differentiation between olive oils against the other oil samples was obtained. In all cases, olive oils tend to be distributed to the right part of the plots across PC1, a component that seems to be related to the polyphenolic content. All other fruit or seed oils apart from olive ones were located to the left part of the PC1 vs. PC2 plot, and a certain discrimination among samples was observed although this was not enough to achieve reasonable clustering according to the fruit/seed of origin (sunflower, soy, or corn). Nevertheless, the results obtained after this exploratory study with PCA were very promising with regard to the authentication of olive oil samples. Regarding the set of 47 EVOO samples, Arbequina and Picual oils were mainly distributed in predominant areas. Oils of non-declared cultivars were basically located in intermediate positions in agreement with their expected blended nature. The HPLC-UV chromatographic fingerprint data was refined by considering only the information of several chromatographic time segments as a way to improve the differentiation and classification of oil samples. For that reason, chromatographic data, at the three wavelengths evaluated, was autoscaled and the following time windows were selected: (1) a 5–8 min segment; (2) an 11–15 min segment; and (3) an 11–20 min segment, as well as the data matrices combining two time windows: (1) 5–8 min + 11–15 min segments; and (2) 5–8 min + 11–20 min segments, and were subjected to PCA. From the direct observation of the HPLC-UV chromatographic profiles obtained from the analyzed samples (see as example Figure S1 in the Supplementary Materials depicting the HPLC-UV fingerprints at 280 nm of two olive oils and a sunflower oil) the studied time segments seem to have, a priori, a certain differentiation, and PCA results showed discrimination between olive oils and other vegetable oils (see two selected examples in Figure S2 in the Supplementary Materials). None of the models evaluated allowed a better classification than the one obtained after considering the full HPLC-UV chromatographic fingerprints (Figure 1).

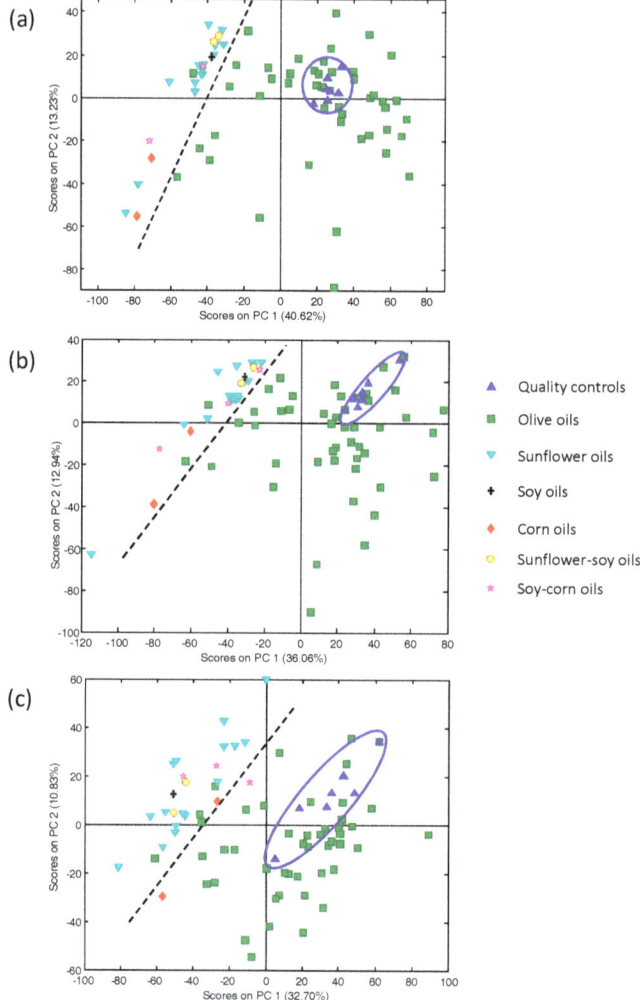

Figure 1. Scatter score plots (PC1 vs. PC2) when processing raw HPLC-UV (High performance liquid chromatography method with ultra-violet detection) chromatographic fingerprints of 47 olive oils and 25 other fruit seed oils (sunflower, corn, soy, and some mixtures of them) registered at (**a**) 257 nm; (**b**) 280 nm; and (**c**) 316 nm. Ellipse is grouping the quality controls.

In a second study, the characterization and classification of EVOOs with respect to the olive cultivar of origin (Arbequina, Picual, Hojiblanca, and Cornicabra) was attempted by using the raw HPLC-UV chromatographic profiles and PCA. For that purpose, a new set of samples was analyzed, consisting of 66 commercially available EVOOs distributed as follows: 23 from Arbequina olives, 19 from Picual olives, 12 from Hojiblanca olives, and 12 from Cornicabra olives. As an example, Figure 2 shows the HPLC-UV chromatographic fingerprints registered at 257 nm from (a) Arbequina; (b) Picual; (c) Hojiblanca; and (d) Cornicabra monovarietal EVOOs. In general, the HPLC-UV chromatographic profiles are different depending on the cultivar, as expected, although some similarities can be observed. These profiles can be divided into three time-window segments. The first one, from 4 to 13 min, is characterized for the presence of very few extracted compounds with low signal intensities. In the second one, from 13 to 21 min, most of the extracted compounds when using ethanol:water (70:30, v/v) solution are eluted. It is characterized for richer profiles with slight differences depending on the olive cultivar of origin. For instance, if we consider some selected time-window sections within this segment, it is possible to see that from 13 to 16 min, a similar signal profile was obtained for the Arbequina, Hojiblanca, and Cornicabra cultivars (with only slight differences in signal intensity) in comparison to the Picual cultivar. From 16 to 17 min it seems that three different signal profiles can be observed: one obtained for Hojiblanca and Cornicabra, and the characteristic ones for Arbequina and Picual, respectively. In contrast, the time-window section from 17 to 21 min clearly provides a very different UV chromatographic fingerprint for Arbequina in comparison to the other three cultivars, which seem to differ only in signal intensity. Finally, a last time-window segment from 23 to 26 min can be considered, the Hojiblanca cultivar being the one showing a characteristic profile in comparison to the other three olive cultivars evaluated. The differences observed in both the chromatographic patterns and the signal relative abundances seem to be representative of each olive variety so that they can be exploited as potential chemical descriptors to achieve classification and authentication of oils by chemometric methods.

Thus, raw HPLC-UV chromatograms of the 66 EVOOs were evaluated as the first PCA model, and the obtained score plots (PC1 vs. PC2) at the three evaluated wavelengths are depicted in Figure S3 (Supplementary Materials). Discrimination of EVOOs regarding the olive cultivar was not achieved successfully under these conditions. Most of the Arbequina EVOOs seem to be clustered together and separated from the other EVOOs at the bottom (data registered at 257 and 280 nm) and at the top (data registered at 316 nm) areas of the score plots, in agreement with the fact that Arbequina EVOOs clearly show the most different raw HPLC-UV chromatographic fingerprints (Figure 2a) in comparison to the other three olive cultivars.

Then, with the aim of improving the classification of EVOOs, several PCA models were evaluated by considering several time-window segments as above. The best results were obtained when HPLC-UV chromatographic profile segments of 5–8 min, 13–21 min, and 23–26 min (registered at 280 nm) were combined. The score plot (PC1 vs. PC2) obtained under these conditions is shown in Figure 3a. As can be seen, by employing only some specific HPLC-UV chromatographic segments, acceptable discrimination and classification among the analyzed EVOOs regarding the olive cultivar of origin was achieved. Quality control samples were clustered together and located at the middle of the model (close to the center area) showing again the good repeatability and robustness of the chromatographic and chemometric results. Discrimination along both PC axes was observed. Most of the Arbequina EVOO samples are located at the bottom-left of the plot, being clearly differentiated through the PC1 and PC2 axes against the other cultivars. Most of the Picual EVOO samples appear grouped at the top of the plot. Hojiblanca and Cornicabra EVOOs are located between the other two olive cultivars, but are still well grouped and differentiated. Obviously, complete discrimination among the analyzed samples is not achieved, as expected. For example, two Picual EVVO samples are clearly located close to where Arbequina EVOOs are clustered, while three Arbequina EVOOs are separated from their group and distributed along the areas assigned to the other three cultivars (see Figure 3a). These samples cannot be considered outliers because this behavior can be expected in this kind of study. It should be taken into account that the total amounts of extracted compounds (mainly polyphenols and phenolic acids), and their distribution in olive oils do not only depend on the olive fruit cultivar of origin but also on many other parameters such as growing area, cultivation techniques, water resources, soil management, degree of olive maturation, climate, and even the oil production method [12–15]. As a result, EVOO samples cultivated or produced under similar conditions might display some similar features and, thus, may be located in close areas. Finally, a last unsupervised PCA was performed by simultaneously combining the HPLC-UV chromatographic fingerprint profiles from 13 to 21 min obtained at the three studied wavelengths (257, 280, and 316 nm). With this model, a data matrix with a dimension of 66 samples × 4053 signals was built and subjected to PCA, and the obtained scatter plot of scores (PC1 vs. PC2) is shown in Figure 3b. The combination of these richer time-window segments obtained simultaneously at different acquisition wavelengths also allowed us to achieve an acceptable classification of the analyzed EVOO with respect to the olive cultivar of origin thanks to the discrimination among both the PC1 and PC2 axes. In this case, quality controls appeared well clustered and close to the center of the plot. Arbequina EVOOs are clustered at the left area of the plot, Picual EVOOs at the top-right section, and Hojiblanca and Cornicabra EVOOs are in the bottom area of the plot and are differentiated through the PC1 axis.

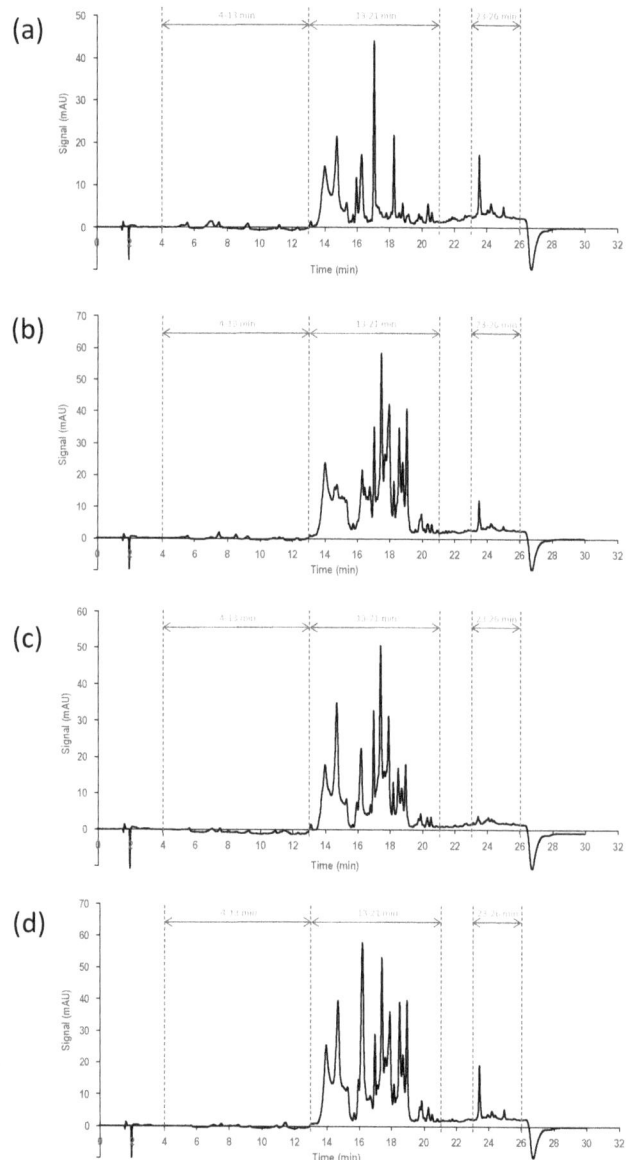

Figure 2. HPLC-UV chromatographic fingerprint registered at 257 nm for four extra-virgin olive oil (EVOOs) obtained from (**a**) Arbequina; (**b**) Picual; (**c**) Hojiblanca; and (**d**) Cornicabra monovarietal olive cultivars.

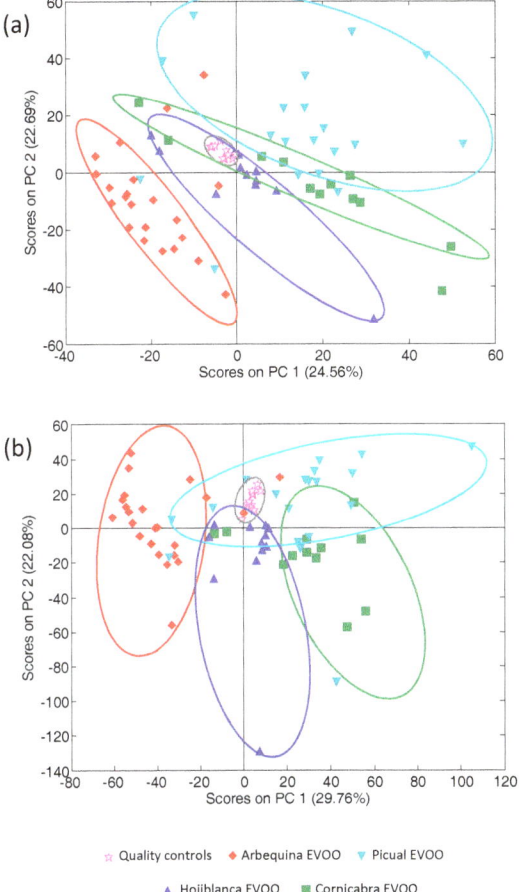

Figure 3. Principal component analysis (PCA) results (scatter score plot of PC1 vs. PC2) in the analysis of 66 EVOOs obtained from monovarietal olive cultivars by employing as data: (**a**) HPLC-UV chromatographic profile segments of 5–8 min, 13–21 min, and 23–26 min simultaneously (registered at 280 nm); and (**b**) a combination of the HPLC-UV chromatographic profile segments from 13 to 32 min obtained at 257, 280, and 316 nm, simultaneously.

The acceptable results obtained from the exploratory PCA of the analyzed olive oils as compared to other fruit seed oils, and the analysis of EVOOs with respect to the olive cultivar of origin, suggested that the proposed HPLC-UV chromatographic fingerprints can be used as potential descriptors to achieve sample authentication and to prevent adulteration fraud in the commercialization of olive oils.

3.2. Authentication and Adulteration Studies by Partial Least Square Regression

The main objective of the present work is to develop a simple and reliable method to confirm the authenticity of EVOOs and to identify and quantify fraud. In the production of EVOOs, product quality is one of the main parameters guaranteeing the olive cultivar of origin, especially when product designations of origin (PDOs) are involved. However, in order to reduce production costs, EVOOs can be adulterated with cheaper fruit seed oils (sunflower, corn, etc.) or even with lower quality olive oils such as refined ones. To study this issue, a specific commercially available Arbequina monovarietal EVOO was adulterated with: (1) a commercially available Picual monovarietal EVOO, in order to evaluate whether the proposed method can guarantee PDOs of monovarietal EVOOs; (2) a commercially available ROO; and (3) a commercially available sunflower oil. In the latter two cases, the focus was on demonstrating the capacity of the proposed method to identify adulteration fraud of EVOOs with cheaper and lower quality oils. Because the difficulty of identifying and quantifying frauds increases when the adulterant is present at low concentrations, adulteration percentages from 2.5 to 85% were studied. The number of samples employed for both calibration and validation sets in the PLS studies is indicated in the experimental section. Thus, adulteration mixtures were prepared and were then analyzed with the proposed HPLC-UV method. The chromatographic fingerprints obtained at the three evaluated wavelengths were subjected to PLS. In all cases, the number of latent variables (LVs) to be used for the assessment of the models was estimated by cross validation based on both Venetian blinds with 5 splits and random blocks (4 blocks and 3 repeats). Results using the two approaches were similar.

For each one of the three adulteration cases studied, the number of LVs was estimated by depicting the latent variable number versus root-mean-square errors in cross validation (RMSECV). Then, the PLS model was built and its performance was evaluated from the scatter plot of actual versus predicted adulteration percentages. Finally, the obtained errors for both calibration and prediction steps were calculated in order to assess the overall quality of the PLS model.

The first PLS model was built from raw HPLC-UV chromatographic fingerprint data without removing any signal at the three UV acquisition wavelengths. Figure 4 shows (a) the estimation of LVs; (b) the validation results in the calibration step; and (c) the validation results in the prediction step obtained in the three adulteration cases studied for HPLC-UV data registered at 257 nm.

As can be seen for the first adulteration study (with a Picual EVOO as adulterant), the lowest prediction error was attained with five LVs, so this number was chosen to quantify the Picual monovarietal percentage with respect to the adulteration of the Arbequina monovariatel EVOO sample. The agreement between the actual adulteration percentages and the predicted ones was highly satisfactory (Figure 4), with calibration errors and prediction errors (summarized in Table 1) of 0.27% and 0.25%, respectively. Similar results were achieved at the other two acquisition wavelengths (Table 1), with overall calibration and prediction errors below 0.67% and 0.41%, respectively. In the second adulteration evaluated (with ROO as the adulterant), the lowest prediction error was also attained with five LVs. Similar to the first case, the agreement between the actual ROO adulteration percentages and the predicted ones was very satisfactory (Figure 4), with calibration and prediction errors (Table 1) of 0.28% and 0.20%, respectively. These errors increased in the less favorable situation only up to 1.54% and 1.42% for the calibration and the prediction models, respectively, when the other two acquisition wavelengths were used. Finally, regarding the third adulteration study (sunflower oil as an adulterant), very similar results as compared to those previously commented were observed, with overall calibration and prediction errors below 0.77% (Table 1) for the three acquisition wavelengths.

Although the results obtained up to this point were very satisfactory, a second PLS model was built by removing from the data matrix the less discriminant HPLC chromatographic segments among the analyzed samples: the death volume segment (from 0 to 3 min) and the chromatographic cleaning and column conditioning segment (from 26 to 30 min). Thus, only the HPLC-UV chromatographic fingerprint from 3 to 26 min (including all the time-window segments discussed in the PCA section) was considered for each of the three adulterations studied. In this case, the lowest prediction error was attained with four LVs, so this number was chosen to quantify adulterant percentages. The corresponding calibration and prediction errors for all the adulteration cases studied are summarized in Table S1 (Supplementary Materials). Again, a very good agreement between the actual adulteration percentages and the predicted values was achieved, with overall errors below 2.88%.

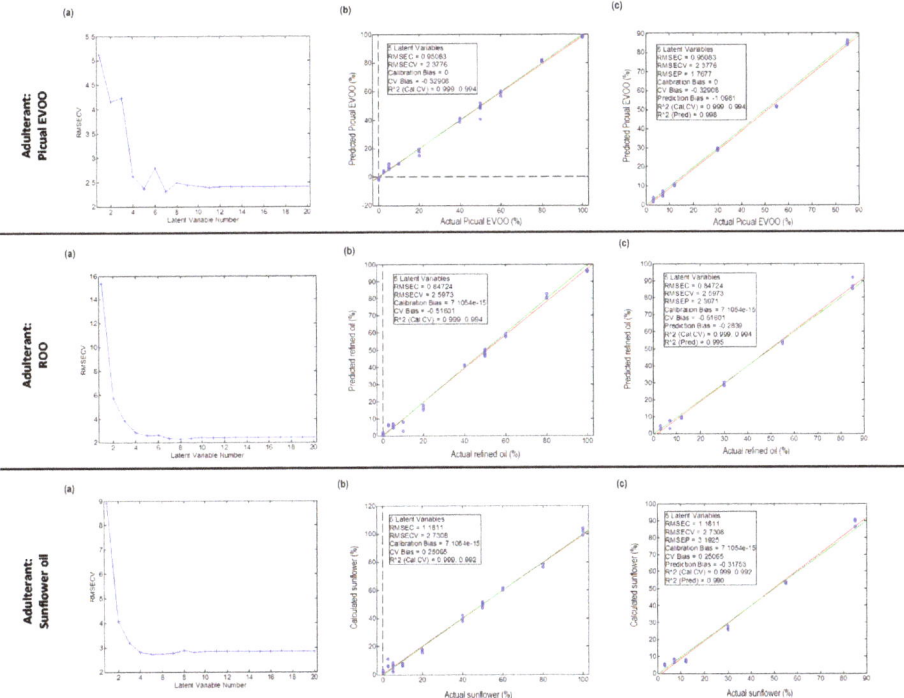

Figure 4. Partial least square (PLS) regression results for the adulteration of an Arbequina monovarietal EVOO using a Picual monovarietal EVOO, a ROO, and a sunflower oil as adulterants. Data set: raw HPLC-UV chromatographic fingerprints registered at 257 nm. (**a**) Estimation of the optimum number of latent variables; (**b**) validation results in the calibration step; and (**c**) validation results in the prediction step. RMSECV: root-mean-square errors in cross validation.

Table 1. Partial least square (PLS) calibration and prediction errors in the identification and quantitation of adulterants (Picual EVOO (extra-virgin olive oil), refined olive oil (ROO), and sunflower oil) in an Arbequina EVOO when using raw HPLC-UV (High performance liquid chromatography method with ultra-violet detection) chromatographic fingerprints as sample descriptors.

Acquisition Wavelength	Calibration Errors (%)		
	Adulterant		
	Picual EVOO	ROO	Sunflower Oil
257 nm	0.27	0.28	0.29
280 nm	0.19	0.33	0.37
316 nm	0.67	1.54	0.77
Acquisition Wavelength	**Prediction Errors (%)**		
	Adulterant		
	Picual EVOO	ROO	Sunflower Oil
257 nm	0.25	0.30	0.54
280 nm	0.41	0.33	0.48
316 nm	0.36	1.42	0.77

Considering the results depicted in Table 1 and Table S1 (Supplementary Materials) we can see that, overall, data acquired at 316 nm produced higher errors (although always quite low and within the accepted parameters) in comparison to the other two acquisition wavelengths. This can be explained by the fact that data at this wavelength seems to be less informative in comparison to the other two, as was previously observed with the QCs employed in the PCA studies (see for instance Figure 1).

After the results obtained in this study, the first PLS model (employing the raw HPLC-UV chromatographic profile) with registered data either at 257 nm or at 280 nm can be proposed to carry out the identification and quantitation of fraud in EVOO samples.

4. Conclusions

A simple, cheap, and reliable method based on HPLC-UV has been developed to characterize and distinguish olive oils (EVOO, VOO, and ROO) from other seed-fruit oils (sunflower, corn, soy), Monovarietal EVOOs were also characterized with respect to the olive cultivar of origin, using chromatographic fingerprints and chemometrics.

Exploratory PCA showed a noticeable discrimination between olive oils and other fruit seed oils (although only slight differentiation within the latter group was achieved) when raw HPLC-UV chromatographic fingerprints (registered at any of the three acquisition wavelengths evaluated) were employed as data descriptors. In contrast, in order to achieve differentiation among monovarietal EVOOs with respect to the olive cultivar of origin (Arbequina, Picual, Hojiblanca, and Cornicabra) the selection of some discriminant time-window segments within the HPLC-UV chromatographic profile was necessary. HPLC-UV chromatographic profile segments of 5–8 min, 13–21 min, and 23–26 min (registered at 280 nm) provided acceptable discrimination among olive cultivars. Similar results were observed when simultaneously combining the HPLC-UV chromatographic fingerprint profiles from 13 to 21 min obtained at the three acquisition wavelengths.

Finally, a PLS study of the authenticity of a specific Arbequina monovarietal EVOO was performed to achieve the identification and quantitation of fraud. The sample was adulterated from 2.5 to 85% by employing three oils as adulterants: (1) Picual monovarietal EVOO; (2) a ROO; and (3) a sunflower oil. A first PLS model employing raw HPLC-UV chromatographic fingerprints without removing any signal allowed the quantification of the studied adulteration frauds with overall quantitation and prediction errors below 1.54%. A second PLS model using only HPLC chromatographic fingerprints within the time segment from 3 to 26 min provided overall quantitation and prediction errors below 2.88%.

Certainly, some adulteration issues can be easily addressed by tracking specific markers or other families of oil components. In any case, the results obtained in this work allow us to propose HPLC-UV fingerprinting and chemometrics as a simple and efficient method for the characterization, classification, and authentication of olive oils. The method is able to guarantee olive oil monovarietal PDOs, and can also detect and quantify EVOO adulteration (minimum 2.5% adulterant) with other less expensive fruit seed oils or cheaper and lower quality olive oils such as refined ones.

Supplementary Materials: The following are available online at http://www.mdpi.com/2304-8158/7/4/44/s1. Table S1: PLS calibration and prediction errors in the identification and quantitation of adulterants (Picual EVOO, ROO, and sunflower oil) in an Arbequina EVOO when using raw HPLC-UV chromatographic fingerprints from 3 to 26 min as sample descriptors. Figure S1: HPLC-UV chromatographic fingerprints registered at 280 nm for olive oils (a,b) and sunflower oil (c). Figure S2: (a) PCA results (scores plot of PC1 vs. PC3) employing as analytical data the HPLC-UV chromatographic fingerprints obtained for time segment 14–20 min at 280 nm; (b) PCA results (score plot of PC1 vs. PC2) employing as analytical data the HPLC-UV chromatographic fingerprints obtained combining time segments of 5–8 min and 11–20 min at 280 nm. Figure S3: PCA results (score plots of PC1 vs. PC2) for the analysis of 66 EVOOs when using raw HPLC-UV chromatographic fingerprints registered at (a) 257 nm; (b) 280 nm; and (c) 316 nm as sample descriptors.

Acknowledgments: The authors gratefully acknowledge the financial support received from Spanish Ministry of Economy and Competitiveness under the projects CTQ2014-56324-C2-1-P and CTQ2015-63968-C2-1-P, and from the Agency for Administration of University and Research Grants (Generalitat de Catalunya, Spain) under the projects 2017 SGR-171 and 2017 SGR-310.

Author Contributions: M.F.-C. and N.C. performed all the experiments. J.S. and O.N. conceived and designed the experiments, supervised the work. M.F.-C. and N.C. wrote the paper.

Conflicts of Interest: The authors declare no conflict of interest.

References

1. *Council Regulation (EC) No 1513/2001 of 23 July 2001 Amending Regulations No 136/66/EEC and (EC) No 1638/98 as Regards the Extension of the Period of Validity of the Aid Scheme and the Quality Strategy for Olive Oil*; Official Journal European Communities: Luxembourg, 2001; Volume L201, pp. 4–7.
2. Rafehi, H.; Ververis, K.; Karagiannis, T.C. Mechanisms of action of phenolic compounds in olive. *J. Diet. Suppl.* **2012**, *9*, 96–109. [CrossRef] [PubMed]
3. Fedeli, E. Lipids of olives. *Prog. Chem. Fats Other Lipids* **1977**, *45*, 57–74. [CrossRef]
4. Waterman, E.; Lockwood, B. Active components and clinical applications of olive oil. *Altern. Med. Rev.* **2007**, *12*, 331–342. [PubMed]
5. Gorinstein, S.; Martin-Belloso, O.; Katrich, E.; Lojek, A.; Číž, M.; Gligelmo-Miguel, N.; Haruenkit, R.; Park, Y.S.; Jung, S.T.; Trakhtenberg, S. Comparison of the contents of the main biochemical compounds and the antioxidant activity of some Spanish olive oils as determined by four different radical scavenging tests. *J. Nutr. Biochem.* **2003**, *14*, 154–159. [CrossRef]
6. Lozano-Sánchez, J.; Segura-Carretero, A.; Menendez, J.A.; Oliveras-Ferraros, C.; Cerretani, L.; Fernández-Gutiérrez, A. Prediction of extra virgin olive oil varieties through their phenolic profile. Potential cytotoxic activity against human breast cancer cells. *J. Agric. Food Chem.* **2010**, *58*, 9942–9955. [CrossRef] [PubMed]
7. Vazquez-Martin, A.; Fernández-Arroyo, S.; Cufí, S.; Oliveras-Ferraros, C.; Lozano-Sánchez, J.; Vellón, L.; Micol, V.; Joven, J.; Segura-Carretero, A.; Menendez, J.A. Phenolic secoiridoids in extra virgin olive oil impede fibrogenic and oncogenic epithelial-to-mesenchymal transition: Extra virgin olive oil as a source of novel antiaging phytochemicals. *Rejuv. Res.* **2012**, *15*, 3–21. [CrossRef] [PubMed]
8. Ricciutelli, M.; Marconi, S.; Boarelli, M.C.; Caprioli, G.; Sagratini, G.; Ballini, R.; Fiorini, D. Olive oil polyphenols: A quantitative method by high-performance liquid-chromatography-diode-array detection for their determination and the assessment of the related health claim. *J. Chromatogr. A* **2017**, *1481*, 53–63. [CrossRef] [PubMed]
9. Hernáez, A.; Remaley, A.T.; Farràs, M.; Fernández-Castillejo, S.; Subirana, I.; Schröder, H.; Fernández-Mampel, M.; Muñoz-Aguayo, D.; Sampson, M.; Solà, R.; et al. Olive oil polyphenols decrease LDL concentrations and LDL atherogenicity in men in a randomized controlled trial. *J. Nutr.* **2015**, *145*, 1692–1697. [CrossRef] [PubMed]
10. Ruiz-del-Castillo, M.L.; Caja, M.M.; Blanch, G.P. Rapid recognition of olive oil adulterated with hazelnut oil by direct analysis of the enantiomeric composition of filbertone. *J. Agric. Food Chem.* **1998**, *46*, 5128–5131. [CrossRef]
11. Moore, J.C.; Spink, J.; Lipp, M. Development and application of a database of food ingredient fraud and economically motivated adulteration from 1980 to 2010. *J. Food Sci.* **2012**, *77*, R118–R126. [CrossRef] [PubMed]
12. Romero, C.; Brenes, M.; Yousfi, K.; García, P.; García, A.; Garrido, A. Effect of cultivar and processing method on the contents of polyphenols in table olives. *J. Agric. Food Chem.* **2004**, *52*, 479–484. [CrossRef] [PubMed]
13. Caponio, F.; Alloggio, V.; Gomes, T. Phenolic compounds of virgin olive oil: Influence of paste preparation techniques. *Food Chem.* **1999**, *64*, 203–209. [CrossRef]
14. Vekiari, S.A.; Koutsaftakisb, A. The effect of different processing stages of olive fruit on the extracted olive oil polyphenol content. *Grasas Y Aceites* **2002**, *53*, 304–308. [CrossRef]
15. Motilva, M.J.; Tovar, M.; Romero, M.; Alegre, S.; Girona, J. Evolution of oil accumulation and polyphenol content in fruits of olive tree (*Olea europaea* L.) related to different irrigation strategies. *Acta Hortic.* **2002**, *586*, 345–348. [CrossRef]
16. Saurina, J.; Sentellas, S. Determination of Phenolic Compounds in Food Matrices: Applications to Characterization and Authentication. In *Fast Liquid Chromatography-Mass Spectrometry Methods in Food and Environmental Analysis*; Núñez, O., Gallart-Ayala, H., Martins, C.P.B., Lucci, P., Eds.; Imperial College Press: London, UK, 2015; pp. 517–547, ISBN 978-1-78326-493-3.

17. Lucci, P.; Saurina, J.; Núñez, O. Trends in LC-MS and LC-HRMS analysis and characterization of polyphenols in food. *TrAC Trends Anal. Chem.* **2017**, *88*, 1–24. [CrossRef]
18. Dais, P.; Boskou, D. Detection and quantification of phenolic compounds in olive oil, olives, and biological fluids. *Detect. Quantif. Phenol. Compd.* **2009**, 55–107. [CrossRef]
19. Segura-Carretero, A.; Carrasco-Pancorbo, A.; Bendini, A.; Cerretani, L.; Fernández-Gutiérrez, A. Analytical determination of polyphenols in olive oil. *Olives Olive Oil Health Dis. Prev.* **2010**, 509–523. [CrossRef]
20. Bonoli, M.; Montanucci, M.; Toschi, T.G.; Lercker, G. Fast separation and determination of tyrosol, hydroxytyrosol and other phenolic compounds in extra-virgin olive oil by capillary zone electrophoresis with ultraviolet-diode array detection. *J. Chromatogr. A* **2003**, *1011*, 163–172. [CrossRef]
21. Farrés-Cebrián, M.; Seró, R.; Saurina, J.; Núñez, O. HPLC-UV polyphenolic profiles in the classification of olive oils and other vegetable oils via principal component analysis. *Separations* **2016**, *3*, 33. [CrossRef]
22. Capriotti, A.L.; Cavaliere, C.; Crescenzi, C.; Foglia, P.; Nescatelli, R.; Samperi, R.; Laganà, A. Comparison of extraction methods for the identification and quantification of polyphenols in virgin olive oil by ultra-HPLC-QToF mass spectrometry. *Food Chem.* **2014**, *158*, 392–400. [CrossRef] [PubMed]
23. Gouvinhas, I.; Machado, J.; Gomes, S.; Lopes, J.; Martins-Lopes, P.; Barros, A.I.R.N.A. Phenolic composition and antioxidant activity of monovarietal and commercial Portuguese olive oils. *JAOCS J. Am. Oil Chem. Soc.* **2014**, *91*, 1197–1203. [CrossRef]
24. Bendini, A.; Bonoli, M.; Cerretani, L.; Biguzzi, B.; Lercker, G.; Gallina-Toschi, T. Liquid-liquid and solid-phase extractions of phenols from virgin olive oil and their separation by chromatographic and electrophoretic methods. *J. Chromatogr. A* **2003**, *985*, 425–433. [CrossRef]
25. García-Villalba, R.; Carrasco-Pancorbo, A.; Zurek, G.; Behrens, M.; Bäßmann, C.; Segura-Carretero, A.; Fernández-Gutiérrez, A. Nano and rapid resolution liquid chromatography-electrospray ionization-time of flight mass spectrometry to identify and quantify phenolic compounds in olive oil. *J. Sep. Sci.* **2010**, *33*, 2069–2078. [CrossRef] [PubMed]
26. Bayram, B.; Esatbeyoglu, T.; Schulze, N.; Ozcelik, B.; Frank, J.; Rimbach, G. Comprehensive analysis of polyphenols in 55 extra virgin olive oils by HPLC-ECD and their correlation with antioxidant activities. *Plant Foods Hum. Nutr.* **2012**, *67*, 326–336. [CrossRef] [PubMed]
27. Alarcón Flores, M.I.; Romero-González, R.; Garrido Frenich, A.; Martínez Vidal, J.L. Analysis of phenolic compounds in olive oil by solid-phase extraction and ultra high performance liquid chromatography-tandem mass spectrometry. *Food Chem.* **2012**, *134*, 2465–2472. [CrossRef] [PubMed]
28. Gosetti, F.; Bolfi, B.; Manfredi, M.; Calabrese, G.; Marengo, E. Determination of eight polyphenols and pantothenic acid in extra-virgin olive oil samples by a simple, fast, high-throughput and sensitive ultra high performance liquid chromatography with tandem mass spectrometry method. *J. Sep. Sci.* **2015**, *38*, 3130–3136. [CrossRef] [PubMed]
29. Carrasco-Pancorbo, A.; Cerretani, L.; Bendini, A.; Segura-Carretero, A.; Gallina-Toschi, T.; Fernández-Gutiérrez, A. Analytical determination of polyphenols in olive oils. *J. Sep. Sci.* **2005**, *28*, 837–858. [CrossRef] [PubMed]
30. Purcaro, G.; Codony, R.; Pizzale, L.; Mariani, C.; Conte, L. Evaluation of total hydroxytyrosol and tyrosol in extra virgin olive oils. *Eur. J. Lipid Sci. Technol.* **2014**, *116*, 805–811. [CrossRef]
31. García, A.; Brenes, M.; García, P.; Romero, C.; Garrido, A. Phenolic content of commercial olive oils. *Eur. Food Res. Technol.* **2003**, *216*, 520–525. [CrossRef]
32. Mastralexi, A.; Nenadis, N.; Tsimidou, M.Z. Addressing analytical requirements to support health claims on "olive oil polyphenols" (EC regulation 432/2012). *J. Agric. Food Chem.* **2014**, *62*, 2459–2461. [CrossRef] [PubMed]
33. Krichene, D.; Taamalli, W.; Daoud, D.; Salvador, M.D.; Fregapane, G.; Zarrouk, M. Phenolic compounds, tocopherols and minor components in virgin olive oils of some Tunisian varieties. *J. Food Biochem.* **2006**, *31*, 179–194. [CrossRef]
34. Garcia, B.; Coelho, J.; Costa, M.; Pinto, J.; Paiva-Martins, F. A simple method for the determination of bioactive antioxidants in virgin olive oils. *J. Sci. Food Agric.* **2013**, *93*, 1727–1732. [CrossRef] [PubMed]
35. Gutiérrez-Rosales, F.; Ríos, J.J.; Gómez-Rey, M.L. Main polyphenols in the bitter taste of virgin olive oil. Structural confirmation by on-line high-performance liquid chromatography electrospray ionization mass spectrometry. *J. Agric. Food Chem.* **2003**, *51*, 6021–6025. [CrossRef]

36. Mazzotti, F.; Benabdelkamel, H.; Di Donna, L.; Maiuolo, L.; Napoli, A.; Sindona, G. Assay of tyrosol and hydroxytyrosol in olive oil by tandem mass spectrometry and isotope dilution method. *Food Chem.* **2012**, *135*, 1006–1010. [CrossRef] [PubMed]
37. Longobardi, F.; Ventrella, A.; Casiello, G.; Sacco, D.; Tasioula-Margari, M.; Kiritsakis, A.K.; Kontominas, M.G. Characterisation of the geographical origin of Western Greek virgin olive oils based on instrumental and multivariate statistical analysis. *Food Chem.* **2012**, *133*, 169–175. [CrossRef]
38. Bajoub, A.; Ajal, E.A.; Fernández-Gutiérrez, A.; Carrasco-Pancorbo, A. Evaluating the potential of phenolic profiles as discriminant features among extra virgin olive oils from Moroccan controlled designations of origin. *Food Res. Int.* **2016**, *84*, 41–51. [CrossRef]
39. Bajoub, A.; Medina-Rodríguez, S.; Gómez-Romero, M.; Ajal, E.A.; Bagur-González, M.G.; Fernández-Gutiérrez, A.; Carrasco-Pancorbo, A. Assessing the varietal origin of extra-virgin olive oil using liquid chromatography fingerprints of phenolic compound, data fusion and chemometrics. *Food Chem.* **2017**, *215*, 245–255. [CrossRef] [PubMed]
40. Gil-Solsona, R.; Raro, M.; Sales, C.; Lacalle, L.; Díaz, R.; Ibáñez, M.; Beltran, J.; Sancho, J.V.; Hernández, F.J. Metabolomic approach for extra virgin olive oil origin discrimination making use of ultra-high performance liquid chromatography-quadrupole time-of-flight mass spectrometry. *Food Control* **2016**, *70*, 350–359. [CrossRef]
41. Eigenvector Research Incorporated. Powerful Resources for Intelligent Data Analysis. Available online: http://www.eigenvector.com/software/solo.htm (accessed on 15 January 2018).
42. Massart, D.L.; Vandeginste, B.G.M.; Buydens, L.M.C.; de Jong, S.; Lewi, P.J.; Smeyers-Verbeke, J. *Handbook of Chemometrics and Qualimetrics*; Elsevier: Amsterdam, The Netherlands, 1997.
43. Pérez-Ráfols, C.; Saurina, J. Liquid chromatographic fingerprints and profiles of polyphenolic compounds applied to the chemometric characterization and classification of beers. *Anal. Methods* **2015**, *7*, 8733–8739. [CrossRef]

© 2018 by the authors. Licensee MDPI, Basel, Switzerland. This article is an open access article distributed under the terms and conditions of the Creative Commons Attribution (CC BY) license (http://creativecommons.org/licenses/by/4.0/).

MDPI
St. Alban-Anlage 66
4052 Basel
Switzerland
Tel. +41 61 683 77 34
Fax +41 61 302 89 18
www.mdpi.com

Foods Editorial Office
E-mail: foods@mdpi.com
www.mdpi.com/journal/foods

www.ingramcontent.com/pod-product-compliance
Lightning Source LLC
LaVergne TN
LVHW070555100526
838202LV00012B/471